THE BRITISH ECOLOGICAL SOCIETY

Ecological Issues Series

Aquaculture: the ecological issues

John Davenport
Kenneth Black
Gavin Burnell
Tom Cross
Sarah Culloty
Suki Ekaratne
Bob Furness
Maire Mulcahy
Helmut Thetmeyer

350 Main Street, Malden, MA 02148-5018, USA
108 Cowley Road, Oxford OX4 1JF, UK
550 Swanston Street, Carlton South, Melbourne, Victoria 3053, Australia
Kurfürstendamm 57, 10707 Berlin, Germany

First published 2003

Library of Congress Cataloging-in-Publication Data has been applied for.

ISBN 1-40511-241-7 (paperback)

A catalogue record for this title is available from the British Library.

Set in Adobe Garamond
by Jackie Bracken, University of Liverpool, and Graphicraft Limited, Hong Kong
Printed and bound in the United Kingdom
by TJ International, Padstow, Cornwall

For further information on
Blackwell Publishing, visit our website:
http://www.blackwellpublishing.com

Contents

Key issues

- Aquaculture is an essential industry providing a crucial part of the world's food supply. Most takes place in low-income, food-deficit countries. It offers the only hope for substantial expansion of aquatic food production in the future, as capture fishery output is stable or declining.

- Aquaculture is very diverse, but dominated in global production and value by kelp, carp, oysters and tiger prawns. Culture of salmon has a high profile in western countries and has attracted criticism for its direct and indirect ecological effects.

- The intensive culture of carnivorous species such as salmon and prawns generally causes greater environmental impacts than more extensive methods employed in the culture of herbivorous species such as carp, oysters and mussels.

- Most forms of aquaculture still depend on harvesting wild resources, either to provide seed, brood stock or food for the cultured species. Seed collection can deplete natural resources or cause damaging by-catch. Aquaculture has tended to 'farm up the food chain' by using industrial fishing products (aquafeeds) to increase production. This intensification of aquaculture increases risks of local pollution and disease transmission, and removes marine food resources that may be important for larger fish, marine mammals and sea birds.

- Considerable damage to coastal ecosystems, particularly coastal mangroves, has happened through aquaculture development, especially in relation to tiger prawn culture. Significant declines in coastal fisheries, and indirect results such as smothering of coral reefs have also occurred.

- Artificial structures used in aquaculture affect local ecosystems. They provide substrata for fouling organisms such as seaweed, barnacles and mussels, and can attract and aggregate fish. Large-scale bivalve farms filter phytoplankton and zooplankton out of the sea, substantially altering local food webs.

- Environmental problems associated with chemical inputs and outputs to aquaculture can be mitigated with present technologies, good regulation and effective management practices.

- Rearing of fish and shellfish can cause stress, which can undermine their resistance, and make them more prone to disease. Movement of fish and shellfish stocks has sometimes resulted in the spread of diseases in the wild.

- Nearly 40% of all known introductions of alien or exotic species to aquatic ecosystems have been related to aquaculture. Some have caused extensive ecological damage.

- The number of inadvertent and deliberate releases of reared aquatic animals to the wild is increasing. Such releases can directly or indirectly reduce the fitness of wild populations and may increase levels of hybridisation with closely related species.

- Aquaculture poses a number of ecological problems. Habitat loss, waste disposal, disease spread to wild populations, genetic pollution and reliance of much of the industry on feedstock derived from capture fisheries are major current concerns. Technological improvements and national/international regulation promise some amelioration, but many aspects of aquaculture as currently practised are ecologically unsustainable.

1 Introduction

Aquaculture has been practiced for centuries, particularly in Asia, but the industry has grown dramatically in the last half century, and has now matured into a major industry in many areas of the world, with numerous training centres and rapid technology transfer to all corners of the globe. In 1998, according to the Food and Agriculture Organisation, aquaculture provided about 26% of global fisheries production and 33% of food fish. Freshwater, marine and brackish water aquaculture of fish and shellfish totals around 30.9 million tonnes per year, compared with 86.3 million tonnes for all capture fisheries, i.e. where the catch is "caught" from wild populations. Aquaculture also provides around 8 million tonnes of aquatic plant material per annum out of a total seaweed production of about 10 million tonnes.

The industry has undoubtedly allowed exploitation of aquatic habitats beyond levels feasible by capture fisheries alone, at least as they are currently managed (Figure 1). It has contributed substantially to maintaining food supplies to rising human populations, and, as it is inherently a rural activity, it often has beneficial knock-on effects on peripheral, fragile economies. Production by capture fisheries is currently controversial, but certainly not increasing significantly. According to FAO statistics it appears to have reached a plateau,

Aquaculture is a major industry of great economic importance

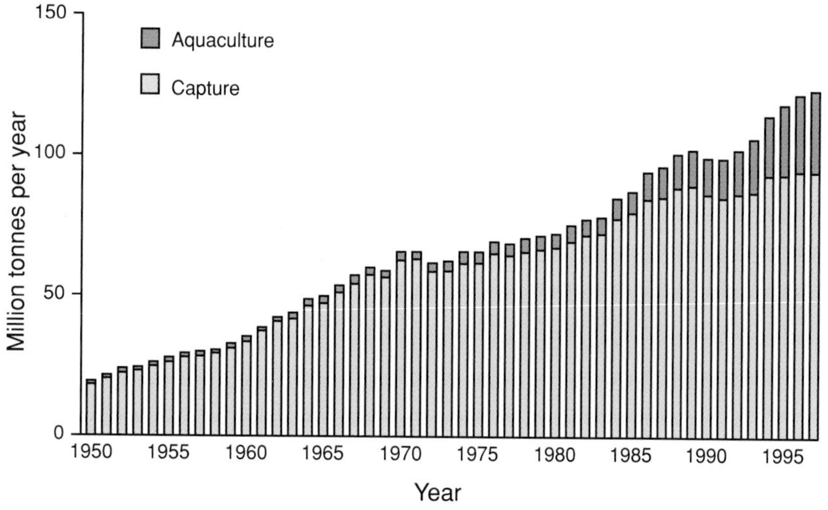

Figure 1. World capture fisheries and aquaculture production between 1950 and 1997. (Source: FAO.)

1

1

Global fisheries are close to their biological limits

Most aquaculture is in low-income food-deficit countries

Most aquaculture is for kelp, carp, oysters and tiger prawns

with growth being no more than 0.6% per year and over 75% of the world fisheries being overfished. However, a recent paper by Watson & Pauly (2001) (strongly disputed by the FAO) indicates that the position may be substantially worse, with a sustained decline in catch of some 660,000 tons per year over the past 14 years. This discrepancy arises because figures supplied to the FAO by China have apparently been regularly distorted, while variability in *Anchevata* catches off South America masks long-term fishery trends. Given either of these scenarios, aquaculture offers the only hope for substantial expansion of aquatic food production in the future, unless management of capture fisheries is significantly improved. It is estimated that aquaculture output will need to reach about 62 million tonnes per year (more than twice the present level) by 2025 to maintain the levels available in 1989 of 19 kg aquatic products per person per year. In 1998 the average person consumed only 15.8 kg, a reduction of 3.2 kg per person, simply because of world population increase.

Growth rates of aquaculture output in many countries have been phenomenal: for example, in Australia the gross value of aquaculture production rose from about $50 million in 1985 to around $400 million in 2000, through the farming of some 60 aquatic species.

While capture fisheries are predominantly marine (90%), aquaculture is still largely a freshwater activity (57%), particularly in the case of finfish such as carp and tilapia. Aquaculture in advanced economies, including Japan, Norway, Canada and the USA, has a high media and scientific profile, but is dwarfed by activity in what the FAO describes as 'low-income food-deficit countries' (LIFDCs). In 1996 around 82% of total finfish, shellfish and aquatic plant production originated in LIFDCs, and production in those countries increased during the early 1990s at six times the rate in non-LIFDCs. Most production is in the Indian subcontinent and the Far East, with China accounting for 83% of LIFDC production. Increasingly these countries rear high-value species including tiger prawns (alternatively known as shrimp) (*Penaeus monodon*) for export to developed economies. These activities make significant contributions to the economies of countries such as Ecuador, Thailand and Indonesia.

The term 'aquaculture' encompasses a great diversity of activities. In freshwater systems around the world about 60 billion fish fry are reared from eggs each year for release into the wild to enhance commercial or recreational fisheries. Herbivorous fish are reared to marketable size in low-technology extensive systems in much of Asia, often in combination with rice farming. At the other extreme, rainbow trout (*Onchorhynchus mykiss*) fed upon artificial pelleted diets are sometimes reared in farms under such crowded tank conditions that liquid oxygen supplies are needed to sustain adequate oxygen levels. Aquaculture includes the rearing of molluscs (abalone, clams, scallops, oysters, mussels), crustaceans and fish for food, as well as the rearing of crocodiles and alligators for luxury handbags. Products can be inexpensive food items such as mussels and clams or internationally stylish items, which include salmon skin wallets and strings of pearls. Specialized aquaculture operations also include the rearing of fancy fish for aquarium hobbyists, bait worms for anglers, laboratory propagation of corals for cancer research, and rearing of the molluscan 'sea hare' *Aplysia* for neurophysiological studies. However, global aquaculture is dominated in volume and value by very few species (Table 1), kelp, carp, oysters and tiger prawns. Worldwide, carp production

1

is the dominant fish-farming activity, while tiger prawns are commercially outstanding because of their high unit value.

Aquaculture was originally regarded as a benign activity. Three decades ago, images of 'farming the seas' or of rice/fish polyculture were viewed as positive when set against the relentless overfishing already shown by many capture fisheries. Aquaculture seemed sustainable, whereas technologically enhanced hunter gathering by fishers did not. However, as the industry has become increasingly competitive and intensive, concerns have arisen, many of ecological significance. During the same period agriculture (the terrestrial model) itself has suffered much justified bad publicity (due for example to overuse of pesticides, subsidy-related habitat degradation and the BSE and Foot and Mouth Disease debacles within the UK) and aquaculture has in turn been subject to more critical examination.

Table 1. The commercial value of world aquaculture production in 1996: the top ten species ranked by volume and value. (Source: FAO, 1998.)

Common name	Species	Production (million tonnes)
Kelp	*Laminaria japonica*	4.17
Pacific cupped oyster	*Crassostrea gigas*	2.92
Silver carp	*Hypophthalmichthys molitrix*	2.88
Grass carp	*Ctenopharyngodon idellus*	2.44
Common carp	*Cyprinus carpio*	1.99
Bighead carp	*Aristichthys nobilis*	1.41
Yesso scallop	*Pecten yessoensis*	1.27
Japanese carpet shell	*Ruditapes phillipinarum*	1.12
Crucian carp	*Carassius carassius*	0.69
Nile tilapia	*Oreochromis niloticus*	0.60

Common name	Species	Production (million US$)
Giant tiger prawn	*Penaeus monodon*	3.93
Pacific cupped oyster	*Crassostrea gigas*	3.23
Silver carp	*Hypophthalmichthys molitrix*	2.79
Kelp	*Laminaria japonica*	2.70
Common carp	*Cyprinus carpio*	2.42
Grass carp	*Ctenopharyngodon idellus*	2.23
Atlantic salmon	*Salmo salar*	1.87
Yesso scallop	*Pecten yessoensis*	1.62
Japanese carpet shell	*Ruditapes phillipinarum*	1.52
Bighead carp	*Aristichthys nobilis*	1.31

1

The open sea is a hostile and demanding environment for construction; most mariculture enterprises have consequently been located in sheltered waters (bays, fjords) or built wholly or partially on land. This leads to conflict with other coastal users. Tropical prawn farming has attracted particular infamy where coastal mangroves have been cleared for prawn ponds and wild postlarvae netted to provide seed for farms. These clearances have sometimes ruined fish nursery areas within mangrove swamps, as well as causing coastal erosion and the destruction of nearby coral reefs. Moreover, removal of postlarvae from the wild may deplete the wild populations of prawns and hence reduce the yield of artisanal capture fisheries.

In developed countries conflicts of a different kind can arise – aquaculture operations are often situated in areas of great natural beauty with attractive, and much valued, wildlife. Coastal managers have to balance the value to residents and tourists of uninterrupted views, with the enhanced prospects of local employment offered by the industry. Aquaculturists have to minimize the depredations of seals and aquatic birds, while respecting environmental laws and changed attitudes to wild predators.

Salmon farming in the northern hemisphere and prawn farming in the south have become increasingly intensive. With this intensification associated disease problems have developed that have damaged the industry economically and triggered fears of disease transmission to wild stocks. Chemical inputs in the form of antifouling agents, anti-fish louse preparations and antibiotics have created situations causing public disquiet. High stock and feeding rates have caused local problems of eutrophication and waste disposal. Because salmon fry, post larval prawns and various cultured molluscs have often been derived from non-native stocks there have been additional concerns over genetic pollution, competition with native species and accidental transmission of pests and diseases.

Another consideration is that of energy input to aquaculture operations. Intensive culture of aquatic plants or herbivorous molluscs and fish sometimes reaches levels limited by the solar input and the primary productivity of the environment. More worrying is the reliance of farming carnivores (e.g. salmonids, prawns) on feed input in the form of pelleted diets. These feedstuffs have been mainly derived from fish meal/oil derived from capture fishing remote from the aquaculture operations. Industrial fishing yielding fish meal and oil accounts for about 30% of capture fisheries production and thereafter requires substantial energy inputs to transform fish into saleable products. Although these products have many uses other than in aquaculture, large-scale fishing for sandeels (*Ammodytes* spp.) that can impact upon seabird populations has been linked with the salmon farming industry in Europe and opened up public debate over the extent to which such aquaculture is truly sustainable.

This booklet does not attempt to describe aquaculture practices in more detail than is necessary, though we indicate sources of such information in the 'Further reading' section.

Instead, we concentrate throughout on known and predictable environmental impacts of aquaculture and their implications for conservation and coastal zone management.

1

Summary

- Aquaculture is an essential industry providing a crucial part of the world's food supply.

- Most aquaculture takes place in low-income, food-deficit countries.

- Aquaculture is very diverse, but dominated in production and value by kelp, carp, oysters and tiger prawns.

- Intensive aquaculture poses a number of problems. Environmental pollution, disease spread to wild populations, genetic pollution and reliance of much of the industry on feedstock derived from capture fisheries are major concerns.

Wild resource harvesting

Unlike terrestrial agriculture, most forms of aquaculture still depend on harvesting wild resources. The life cycle of cultured aquatic species is often complex or incompletely understood. Even if the requirements for successful reproduction are known (as in penaeid prawns for example), it may be technologically demanding or economically impracticable to produce sufficient quantities of the required seed under controlled conditions. Thus, many aquaculture practices involve stocking with wild seed, such as eggs, larvae, spat and fingerlings (juvenile fish) or the capture of brood stock from the natural habitat.

The cultured species may depend completely or partly on natural food resources. In some cases, the farmed animals collect wild food by themselves, for example in mussel culture or in extensive pond culture, or are fed with an artificial feed containing fishery products such as fish meal and fish oil.

Much aquaculture relies on capture of wild 'seed', so is not closed-cycle

2.1 Seed collection

Wild seed of aquatic organisms continue to be captured from the wild for a variety of organisms, though the development of breeding technologies and their transfer to Developing Countries has meant that more countries are now adopting hatchery practices to produce seed for culture.

The obvious benefits are that such hatchery practices bestow the advantages of making stocking cycles independent of natural breeding times and that improved seed with desirable characteristics can be produced in hatcheries for farming operations. Even so, wild seed continue to form the basis of culture for species where breeding in captivity has still not been developed to commercially broadbased levels (as in the case of milkfish, *Chanos*), or in countries where hatchery production of seed is as yet insufficiently developed to meet culture demand (as in the case of tiger prawns in Bangladesh).

Sometimes, even where breeding may be possible, technology for larval culture may not have advanced sufficiently to make hatchery rearing economically feasible in practice. For example, in groupers, it is reported that China and Taiwan have had recent success in the hatchery rearing of estuarine groupers, *Epinephalus coioides* and *E. malabaricus*. However, these methods, developed in private sector hatcheries, are not generally disseminated and the price of hatchery reared juveniles remains high so that the main source of grouper juveniles that are caught in many Asian countries continues to be the wild.

Wild collection of fry and fingerlings target specific species and their specific sizes. However, capture methods have not been developed to be sufficiently specific to exclude other organisms that are therefore also captured as by-catch. This by-catch is often discarded and is an added ecological cost in the wild collection of seed. In Bangladesh, the

2

capture of a single tiger prawn post-larva is accompanied by the indiscriminate capture and destruction of some 1400 macro-zooplanktonic individuals, including other prawn post-larvae and fish larvae.

In the wild collection of seed, it is not only the destruction of by-catch organisms that has an ecological impact, but also the destruction of the habitat itself. This is because damaging methods are often used to catch seed. Push-nets and some types of cast-nets that are used in estuarine and mangrove bottoms to catch prawn and grouper seed alter or destroy the bottom habitats that often serve as nursery beds.

Even where hatchery practices are developed for some species, cheaper wild-caught seed sometimes fetch sufficiently high prices to warrant their being caught from the wild, especially when unemployed or low income communities in tropical developing countries are involved in their capture. In this way, the uncontrolled capture of grouper seed that were exported to South East Asian countries from Sri Lanka destroyed bottom habitats as well as reducing grouper numbers severely. A drastic reduction in wild grouper juvenile numbers has also been reported from Hong Kong and China.

Although collection of estuarine grouper juveniles is also practiced using an environmentally acceptable method using aggregation devices made of materials such as mangrove brush in Sri Lanka and Philippines, reduced catches have been reported from the Philippines through shallowing of estuaries due to sedimentation as well as water quality and circulation changes resulting from uncontrolled proliferation of fish ponds.

Capture of wild seed results in ecological costs

Mussels

Mussel farming often relies on the harvest of wild mussel spat to establish cultures of mussels on ropes or on selected intertidal or subtidal plots. Removal of natural spat can be very extensive and can influence food availability for specialist predators of mussels. It is difficult to estimate the importance of mussel spat harvest for dependent wildlife. The Irish and Dutch dredged mussel fisheries rely upon the transplantation of both inter-tidal and offshore seed to sustain them.

The sub-tidal seedbeds are often ephemeral, being swept away by winter storms. It might therefore be argued that the inter-tidal beds should be kept for the birds and the sub-tidal beds exploited by the mussel dredgers. However this simplistic argument ignores the fact that the best settlement surface for planktonic mussel larvae appears to consist of existing mussel beds! Any management plan that attempts to integrate the needs of man, birds and mussels must take this fact into account.

In many areas the quantity of spat is so large that harvesting some may have little or no influence on subsequent biomass of natural mussel stocks in the region. However, there is one well documented case where bird populations have been severely affected by mussel spat harvest and mussel farming, combined with fisheries for other molluscs that might have provided alternative prey. In the Wadden Sea, eider ducks and oystercatchers occur in hundreds of thousands during winter, and feed predominantly on mussels and cockles. These resources are subject to heavy fishery harvest as well as collection of mussel spat for subtidal mussel farming, despite the fact that the Dutch Wadden Sea has been declared a wetland of international importance as a RAMSAR site, a Biosphere Reserve, and under the EC Wild Birds Directive and EC Habitats Directive.

In 1990, mussel seed collection led to a near-complete stock depletion, at the same time as harvesting of mature cultivated mussels reduced adult stocks to a very low level. Unprecedented thousands of eiders and oystercatchers died that winter and numbers in the Wadden Sea fell and have remained lower than in the 1970s and 1980s. Many surviving eiders moved out of the Wadden Sea and established a new pattern of feeding on another mollusc species, *Spisula*, available in deeper North Sea waters. Continued harvesting of mussel spat and contraction of the area of mussel beds into smaller areas of protected mussel farms, together with fisheries exploiting Wadden Sea cockles and North Sea *Spisula*, was followed in winter 1999/2000 by another mass mortality of eiders, with over 21,000 birds starving to death along the Wadden Sea coastline.

2.2 Brood stock collection

Lobsters
European lobster (*Homarus gammarus*) catches have declined from circa 3,500 tonnes per year in the 1930s to less than 2,000 tonnes per year in the early 2000s. As they are one of the most valuable of shellfish this has prompted much research into the possibility of their culture. At the moment the only economic way to obtain juveniles is to capture females with eggs and hold them in hatcheries until they release their larvae. The larvae are then reared up to the benthic phase when they can either be reared in individual chambers to avoid cannibalism, or released back into the wild. Some recent release experiments in Norway and the UK have given encouraging returns of up to 9% market size

lobsters within 5-6 years. However, the success is very dependent upon the nature of the release site, with a large cobble sized substrate giving the best results.

Although it is possible to carry out controlled mating in captivity, it is difficult to achieve predictable results on a routine basis and so commercial hatcheries are relying on wild brood stock that has already mated in the wild. The captured females with extruded eggs are referred to as 'berried' and they may have to be held for several months at an elevated temperature before the larvae hatch and are released. Release trials in both Norway and Italy (in the northern Adriatic) have been promising and the idea is to restore local stocks by managing the fishery through fishery cooperatives. Because lobster rearing is relatively small scale at the moment this practice is not a problem. However, in the future there could be such a demand for berried females that wild recruitment could be compromised. It is therefore important that more research is carried out into controlling the whole life cycle in captivity.

Penaeid prawns
Penaeid prawn pond culture accounted for 30% of total world penaeid supply in 1998 and production continues to rise. Prawn farms in much of the world still rely on the collection of wild larvae for stocking into ponds. This causes conflicts with prawn fisheries and limits the areas in which prawn farming can be conducted – they have to be near the source of wild seed. Over the last 20 years prawn hatcheries have developed, with adult brood stock taken from the wild and induced to spawn in captivity. Techniques for captive maturation and spawning have advanced considerably, but fundamental understanding of ovarian development in *Penaeus* sp. is still

Capture of seed mussels can affect bird populations

2

lacking, and the details of fertilization processes are obscure. Partly because of this, the prawn hatchery industry finds that wild-matured brood stock generally produce superior penaeid nauplii and postlarvae to those derived from either captive-matured or captive-bred adult prawns. The recent spread of disease in the tiger prawn farming industry no longer permits the use of brood stock from previously-fished nearshore areas. To avoid collecting diseased brood stock, previously unexploited offshore areas are now being targeted and use of captive-bred stock is declining.

There is now a thriving international trade in adult prawn brood stock (wild or captive-bred) as well as the long-established trade in prawn eggs and postlarvae. *Penaeus japonicus* have been exported from Japan to both Europe and South America, while *Penaeus monodon* from SE Asia have been exported to almost all tropical countries, since it is the most valuable of penaeids. The trade has largely been unregulated and has resulted in frequent transmission of pathogens, causing massive commercial losses. It is likely that it has also resulted in escapes of non-native species or subpopulations into many areas, though evidence of significant consequences is lacking.

Movement of broodstock can transfer pathogens to wild populations

The serious consequences of transmission of pathogens is likely to lead to an increasing concentration on use of postlarvae derived from captive brood stock certified to be disease-free. This will take the pressure off wild populations and facilitate breeding of disease-resistant strains. However, genetic diversity in captive-bred stock is known to be lower than in wild populations; this may lead to problems if escapees prove to be a problem in the future (see Section 6).

2.3 Feedstock

Aquafeed

Industrial fishing, to harvest small fish for production of artificially-compounded feed for aquaculture ('aquafeed'), is an indirect, but important, influence of the aquaculture industry on the environment. Not only has aquaculture production been increasing, but there has also been a trend towards intensification of aquaculture systems. A particularly rapid increase has occurred in the use of aquafeed in production of predatory fish and crustacea.

Aquaculture has tended to 'farm up the food chain'. This trend has been most pronounced in salmonid culture in developed countries and prawn culture, much less so in staple freshwater fish culture in countries such as the USA China, Indonesia, the Philippines and India – though some of these already farm prawns and are moving into luxury marine fish culture with pelleted diets.

Globally, the use of aquafeeds to increase production of farmed finfish and crustacea has increased at a rate in excess of 30% per year for the last few years. Production of around 3 million tonnes of farmed fish and crustacea in 1995 required about 1.5 million tonnes of fish meal and oil, manufactured from over 5 million tonnes of pelagic fish (wet weight). By 2000, nearly 10 million tonnes of pelagic fish was being used in aquafeeds, and this increase is projected to continue.

These pelagic fish are harvested by 'industrial fisheries', where the entire catch is destined for reduction to fishmeal and oil rather than for direct human consumption. Most industrial fisheries harvest from stocks of abundant,

Much aquaculture relies on fishmeal diets, and hence on capture fisheries

small, shoaling, pelagic fish such as anchovies, sandeels, capelin, sprats or juvenile herring. Such fish are also food for many predatory fish that are commercially important for human consumption fisheries, as well as supporting many marine mammals and seabirds. Hence concern has been raised over potential for competition between industrial fisheries, commercially important human consumption fisheries, and wildlife conservation interests.

Fish-based aquafeeds are essential in culture of predatory fish, as those fish require long-chain n-3 polyunsaturated fatty acids in their diet. However, fish-based aquafeeds are also beneficial in herbivorous fish production. There are several compelling arguments in favour of using aquafeeds based largely on fishmeal and fish oils. When fed partly on fishmeal and oil, herbivorous fish achieve higher growth rates. They also benefit from a more robust immune system due to the omega-3 rich fish oil and so may survive stresses of intensive aquaculture better. Furthermore, using feeds rich in n-3 fatty acids permits production of farmed fish with a high content

of these fatty acids, so provides beneficial health effects for human consumers.

Aquafeeds normally contain large quantities of fishmeal and fish oil derived from the catches of 'industrial fisheries'. However, industrial fisheries would exist even in the absence of aquaculture, since fishmeal and oils are important constituents of feeds for chickens, pigs and other animals in intensive agriculture; in 1994 about half the world production of fish meal was fed to chickens (Figure 2).

Meal and oil from industrial fisheries forms only a small component of the feed of intensively reared pigs and chickens, but they are the main constituents of aquafeeds, so that up to six tonnes of industrial fish is needed in the production of one tonne of cultured marine fish (Figure 3). Furthermore, the proportion of the global catch of industrial fisheries that is used for aquafeeds has been increasing very rapidly, from 10% in 1988 to 17% in 1994, and 33% in 1997.

This gives several causes for concern. The rapid increase in requirement for fishmeal and oils

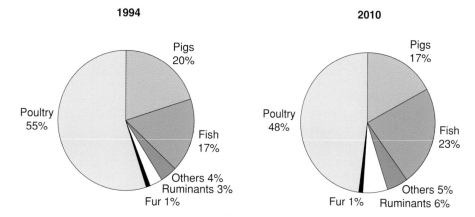

Figure 2. Estimates of the use of fishmeal in1994 and 2010. Data suggest a 3% increase in fishmeal use during this period. (Source: data of fishmeal trade association IFFO (http://www.iffo.org.uk/).)

11

2

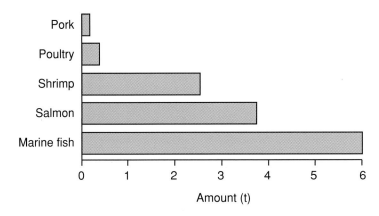

Figure 3. The amount of industrial fish (tonnes) used to produce one tonne of farmed animals.

could not be met by proportional increases in industrial fish landings, since most major industrial fisheries are fully exploited. By comparison with relatively slowly increasing industrial fish catches, global production of soya has doubled in the last ten years. This has led to increasing prices for fishmeal and fish oil whereas soya meal and oil prices have remained relatively constant (Figure 4); this provides an increasing incentive to replace fish with soya in feeds where this is possible.

Even the current levels of industrial fishing may not be sustainable. For example, Peruvian anchovy catches drop considerably in El Niño years. North Sea sandeel (*Ammodytes* spp.) catches may only be sustainable as long as stocks of predatory fish remain greatly depleted. It is believed that capelin (*Mallotus villosus*) catches may only remain high while cod stocks that feed on those capelin remain depleted. Collapse of the Barents Sea capelin stock during the 1980s seems to have been due

Six tonnes of industrial fish are needed to raise one tonne of cultured fish

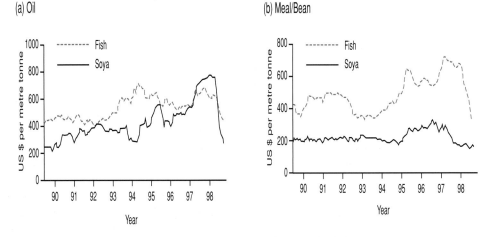

Figure 4. Prices of fish meal and oil compared with soya meal and oil. (Source FAO.)

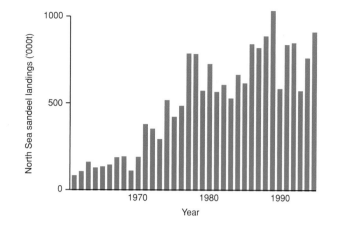

Figure 5. Sandeel landings from the North Sea each year from 1961 to 1995.

predominantly to very high capelin mortality as a result of rapid increases in cod recruitment, but coupled with an additional impact of industrial fishing on the capelin stock.

Because industrial fishery catches can be very large, it has been argued that industrial fisheries may have adverse effects on the rest of the food web of which the industrial stocks are a fundamental component. For example, the North Sea sandeel fishery has grown very rapidly over the last three decades to become the largest single species fishery in the North Sea (Figure 5).

Industrial fisheries might deprive top predators of their food, or may alter energy flow within a food web to change the balance between fish stocks. Interactions between industrial fisheries and other parts of the ecosystem have been most intensively studied in the North Sea (sandeel fishery) and in the Barents Sea (capelin fishery).

In the case of North Sea sandeels, many seabirds, marine mammals and predatory fish feed predominantly on sandeels in summer when these fish become available in the upper layers of the sea. We know much more about seabird feeding on sandeels than about consumption by marine mammals or predatory fish. Ecological theory predicts that some types of seabirds will be much more vulnerable to reductions in food-fish abundance than others. Empirical data from Shetland, where sandeel abundance fell to dramatically low levels in the mid-1980s, support these theoretical predictions.

One of the species expected to respond most strongly is the kittiwake (*Rissa tridactyla*). Breeding success of kittiwakes does indeed correlate with sandeel stock density, both in the North Sea, and in Shetland (Figure 6). The North Sea sandeel fishery was closed from January 2000 in one small area of the North Sea where kittiwake-breeding success had been poor for several seasons following large harvests of sandeels, on the basis that kittiwake performance represents an indicator of the availability of sandeels to top predators in general. However, kittiwake-breeding success has been high in most North Sea colonies, and breeding numbers have increased alongside the growing industrial fishery. Why has this been possible? It seems that reductions in stocks of predatory fish have more than compensated for

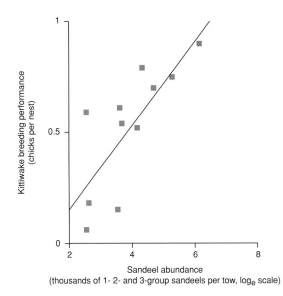

Figure 6. Breeding performance of kittiwakes in relation to numbers of sandeels near Shetland.

the growing industrial fishery. Predatory fish, especially mackerel and gadoids, are by far the largest consumers of sandeels, while the needs of seabirds and marine mammals are very much less (Figure 7).

The implication of this is that changes in abundance of predatory fish probably influence sandeel availability to seabirds and seals more than the industrial fishery does. A large industrial fishery may be compatible with healthy populations of seabirds and marine mammals in the North Sea providing predatory fish stocks remain depleted.

In the Barents Sea, stocks of cod and herring have been heavily exploited for human food consumption, and capelin are harvested by an industrial fishery. The capelin stock collapsed in the mid-1980s and again in the early 1990s. These collapses were largely due to high predation rates by cod, resulting from

exceptionally large cod recruitment in certain years, but exacerbated by the industrial fishery for capelin.

The consequence for seabirds and marine mammals was dramatic. For example, huge numbers of starving arctic seals invaded Norwegian coastal waters in search of food. Almost 90% of common guillemots (*Uria aalge*) in the Barents Sea starved to death in winter 1986-87 because they could not find alternative food in the absence of capelin.

Despite these catastrophes, there are common features between the Barents Sea and North Sea ecosystems. In both, predatory fish are by far the biggest consumers of the food fish, with the industrial fishery and marine mammals taking less, and seabirds less again. In both systems, fluctuations in predatory fish stocks appear to influence food fish availability to wildlife more than the industrial fishery does. However, both

2

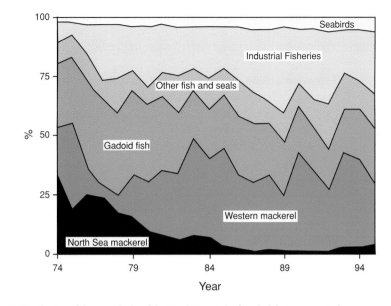

Figure 7. Distribution of the annual take of the North Sea stock of sandeels between particular consumer groups; North Sea mackerel stock, east Atlantic mackerel stock known as 'western mackerel', gadoids (primarily whiting, haddock and cod), all other piscivorous fish plus seals, the North Sea industrial fishery, and seabirds. Quantities were estimated by bioenergetics modelling plus data on stomach contents of fish and diet composition of seabirds and seals.

Industrial fishing can adversely affect populations of seabirds

cases show that top predators such as seabirds and marine mammals are vulnerable to alterations in ecosystem structure, whether induced by natural variation or by fisheries.

Elsewhere, interactions between industrial fisheries, marine mammals and seabirds have not been quantified in detail. However, the loss of millions of seabirds from the Peruvian coast is a well-known example where the industrial fishery for *Anchoveta* seems to have reduced food availability to the birds, thereby inhibiting their populations from recovering after crashes induced by El Niño events. Although the influence of food-fish abundance on the recruitment and growth rate of predatory fish is not well understood, it is possible that industrial fisheries may inhibit the recovery of depleted predatory fish stocks by reducing densities of their food-fish.

Soya is increasingly substituting for fish meal in animal feed – but overall demand for fish meal still increases

A trend in aquaculture has been to reduce the quantity of fishmeal and oil used to produce a unit of aquaculture produce. Improvements in the feeding efficiency of aquaculture systems have been taking place, so that the amount of feed wasted in intensive systems has been reduced. This is important in reducing farm waste pollution impacts (see Section 4), but higher conversion efficiency also reduces the need for industrial fishing products.

At present, most aquafeeds are over-formulated as nutritionally complete diets which take no account of stocking density and natural food availability, and this requires adjustment to further reduce wastage. Another trend that may mitigate increased requirements for aquafeeds is the alteration of aquafeed composition to contain less fishmeal and oil. Industrial fish protein can to some extent be substituted by

2

soya protein or by the use of discards and offal from human consumption fisheries (this would be an elegant way to reduce the discard problem as well as the industrial fishing problem). Fish oil can similarly be substituted by vegetable oil or by oil from discards and offal. Specific limiting nutrients can be added to try to compensate for chemical compositional differences between fish and other ingredients.

At the moment there is much research by commercial aquafeed companies to investigate these alternatives and the suitability of altered feeds for aquaculture production and product quality. IFFO (a trade association) has estimated that use of aquafeed will more than double between 2000 and 2010 (Table 2). Despite reductions in the proportions of

fishmeal and oil in aquafeed, the quantities of these constituents needed to support this growth are anticipated to increase by about 30% over that decade (Table 3).

Although FAO estimate that about 25 million tonnes of fish are discarded worldwide each year, mechanisms and facilities to collect these and to convert them to fishmeal and oil are rarely available. Furthermore, many contaminants are stored in fish livers and accumulate to higher concentrations in larger and older fish, so contaminant problems might arise if discards and offal from continental shelf and enclosed sea fisheries were substantial contributors to fishmeal. Most industrial fish are short lived and occur in upwelling regions where pollution is negligible. While it seems that some changes that reduce dependence on

Table 2. Projections of aquaculture production by species groups and estimated requirements for aquafeeds.
(Source: data of fishmeal trade association IFFO.)

Species	Estimated production in 2000 ('000 t)	Predicted production in 2010 ('000 t)	Estimated % to be reared on aquafeed in 2000	Projected % to be reared on aquafeed in 2010	Aquafeed required in 2000 ('000 t)	Aquafeed required in 2010 ('000 t)
Carp	13,983	36,268	25	50	6,991	27,000
Tilapia	974	2,526	40	60	779	2,106
Prawns	1,034	1,684	80	90	1,489	2,425
Salmon	876	1,569	100	100	1,051	1,255
Bass, etc	856	1,394	60	80	923	1,670
Trout	450	733	100	100	585	586
Catfish	371	604	85	90	505	761
Milkfish	379	462	40	75	303	554
Eel	216	263	80	90	346	284
Other marine fish	105	650	100	100	126	585
Total	19,244	46,153			13,098	37,226

Table 3. Predicted requirements ('000 t) for fishmeal and oil for use in aquafeeds in 2010 compared with present use in 2000. (Data from IFFO.)

Species	Fishmeal		Fish oil	
	2000	**2010**	**2000**	**2010**
Carp	350	675	70	135
Tilapia	55	74	8	11
Prawns	372	485	29	73
Salmon	454	377	283	251
Bass, etc	415	668	185	251
Trout	176	147	88	1177
Catfish	15	-	5	8
Milkfish	36	28	10	11
Eel	173	114	17	28
Other marine fish	69	263	13	70
Total	2,115	2,831	708	955

Brine shrimp larvae are important food items for young fish and prawns – but demand exceeds supply

fish meal and oils may be practical, it is likely that the overall trend for the foreseeable future will be greater demand for fish meal and oil in aquaculture rather than a reduction, and it is probable that inclusion of discards and offal will represent only a small fraction of future aquafeed production.

Artemia

The brine shrimp *Artemia* is crucial to the rearing of a great range of fish and crustaceans. *Artemia* is always collected from the wild, commonly as cysts but sometimes as adults. The latter are block-frozen and sold to hatcheries and farms. Brine shrimps live in salt lakes or salterns (man-made salt pans) and are part of the low diversity ecosystems that are compatible with fluctuating salinities and periods of extremely high salt concentrations. *Artemia* cysts (eggs) are collected from the lakes, cleaned, dried and vacuum packed. They can be stored almost indefinitely in freezers and hatched whenever convenient.

The young stages of cultured aquatic animals often depend on small-sized live prey. Brine shrimp nauplii (larvae) can easily be produced from canned cysts and therefore have become the most popular live prey in aquaculture hatcheries (e.g. for rearing of flatfish, prawns, and lobsters). The poor nutritional value of *Artemia* has been overcome by feeding (enriching) the newly hatched nauplii with emulsions containing essential fatty acids, free amino acids, or vitamins.

Roughly speaking the worldwide *Artemia* market between 1999 and 2000 is about 6,200 tonnes (wet mass) of *Artemia* cysts per year (about four hundred thousand billion cysts!). In the last 3-5 years a severe shortage of *Artemia* has developed as demand has outstripped supply. The most prominent source of *Artemia* for many years has been the Great Salt Lake in Utah, USA.

It is now evident that overexploitation of the

2

resource has taken place. In most years around 85% of cysts have been collected each autumn, while modelling suggests that 5-20% of cysts need to be left in the lake to restart *Artemia* production in the subsequent spring. In October 1997, the Department of Natural Resources in the State of Utah, halted the Great Lake harvest several months early after noting raised algal levels and declining numbers of adult brine shrimp, females with eggs, numbers of eggs per female and numbers of free cysts in the lake during September 1997. This caused severe shortages and elevated prices which was ameliorated to some extent by the introduction of *Artemia* supplies from Russia. Reduced harvests have continued in 1998-1999. The major *Artemia* suppliers are now unable to guarantee supply and are exhorting customers to consider alternative diets. In the longer term it is evident first that *Artemia* stocks will require careful management, and second that predicted future expansion of the aquaculture industry cannot rely on increased exploitation of brine shrimps.

Summary

- Most forms of aquaculture still depend on harvesting wild resources.

- Seed collection often involves damaging by-catch – capture of a single tiger prawn post-larva can involve destruction of 1,400 planktonic animals of similar size.

- Industrial fishing for small fish that are converted into aquafeed has tended to 'farm up the food chain' and removes food resources for larger fish, marine mammals and sea birds.

- Up to 6 tonnes of industrial fish are needed for the production of 1 tonne of cultured fish.

3 Physical change to the habitat

Aquaculture practices, other than in ranching, often involve physical changes to the culture habitat to yield the necessary holding facilities and related infrastructure. As culture operations change from small-scale subsistence culture to large-scale extensive or intensive culture, the physical changes that are involved become more and more permanent and make increasingly higher transformation. The physical changes and disturbance to habitats were justified in the early days of culture operations such as prawn rearing by the view that this involved the implanting of a new industry onto otherwise unproductive rural land, otherwise unsuited for profit-making land-use activities such as agriculture. This was because coastal wetlands in particular were then considered as 'marginal' or 'wasteland' and were used more as dumping grounds because their valuable ecological functions or services were not well established.

With later studies establishing, and progressively increasing, our understanding of the unique ecological functions of wetlands, a precautionary approach has been developing. More recently, the physical impacts of aquaculture on coastal wetlands have received serious concern and there has been general agreement that there should be no further loss of fragile ecosystems, such as mangroves and other wetlands. Unfortunately this has not been properly implemented in many developing countries.

The construction of holding facilities such as ponds as well as infrastructure facilities such as roadways, staff housing, power and communications pylons transform terrestrial physical habitats, and alongside the construction of holding facilities, increase the area of land used much beyond the area of the actual farm. Hatchery construction is often less damaging when it is on high land.

Physical habitat changes of a semi-permanent nature can be brought about directly, by constructing holding structures for farmed stock such as ponds, or indirectly by unsustainable seed collection activities (see Section 2.1). Other indirect changes include the salinisation of fresh waters, and the alteration of sediment loads in neighbouring aquatic systems. Some of these consequences are explored further below.

3.1 Creation of pond wetlands

The conversion of terrestrial and semi-terrestrial habitats into ponds has a major direct impact on ecosystems. In the tropics, many types of habitats have been transformed into pond wetlands. Wetlands such as coastal mangroves (discussed in detail in Section 3.2) and marshland, as well as inland agricultural land that include coconut, paddy, sugar, sparsely used croplands, have undergone change in this manner.

The creation of brackish water pond wetlands for prawn culture has brought about unanticipated changes to surrounding habitats, such as the salinisation of soils and waterways. Once pond wetlands are created by physically

3

altering the habitats, the availability of brackish water of acceptable quality becomes one of the greatest constraints to prawn culture. Therefore, when inland habitats are converted to brackish ponds, seawater has to be brought in and pumped in to the ponds to bring up salinities to the desired levels. Seepage of salt-water into surrounding habitats then takes place. Such seepage from prawn ponds into feeder channels of fresh water habitats and paddy land has often been found. For example, in Sri Lanka, salt-water intrusions have been experienced up to 6 km upstream in paddy growing areas and have damaged paddy cultivation. While this affects the socio-economic well being of dependent traditional communities, such changes to habitats also affect the biodiversity of these once fresh-water areas, changing the diversity and abundance of faunal assemblages.

In the dry season, maintaining pond water salinities of these converted brackish habitats causes an opposite environmental impact. With increased evaporation in the dry season, salinities rise and it becomes necessary to use fresh water to dilute the pond waters. Usually groundwater is used. Excessive extraction of groundwater to reduce these increasing pond water salinities brings about salt water intrusions into fresh drinking and agricultural water and even to land subsidence, as has happened in Taiwan.

These increases in dry-season salinities have been more common in recent years; for example, average annual salinities have risen by 25% over the last thirty years (from 36 to 46%) in lagoons used for prawn farming on the west coast of Sri Lanka. Decreased freshwater inputs, flow reductions through increased farm use, and physical obstructions to flow regimes by farm constructions are thought to cause

these progressive increases in salinities. Although widely discussed, it is yet to be established whether wider processes such as the changing global climatic patterns and related global warming are also involved in the progressive salinity build-ups. These processes have already destroyed over 80% of coral reefs in several tropical countries.

While these changes often alter the biology, ecology and hydrology of habitats irreversibly, they have also led to much social disarray including land use conflicts and changes to traditional practices in agriculture, land use, animal husbandry and fishing. Even when prawn pond activities cease and the ponds become abandoned (see below for habitat alteration resulting from abandonment of prawn ponds), the salination of ground water and salinisation of soils prevents their reconversion to agriculture for many years.

The creation of pond wetlands triggers further habitat changes due to the continuous activities that go on in the farms. For example, effluents discharging from farm ponds and hatcheries have considerable impacts when the discharges are confined to water bodies such as estuaries. Poorly managed farm effluents increase eutrophication, toxic algal blooms, disease transfer between farms and hatcheries and the accumulation of antibiotic-resistant bacteria. In the receiving waters, they are known to have increased salinity, turbidity, total suspended solids and BOD (Biological Oxygen Demand), toxic metabolites such as phosphates, nitrates, nitrites, sulphides, and heavy silt loads leading to decreased depth and water circulation. These aspects are dealt with in detail later (see Section 4).

More indirect effects of the creation of pond wetlands have been seen by an increased

Serious damage had been caused to coastal wetland by pond culture of fish and prawns

Irreversible salination of soils has resulted from mangrove clearance for pond culture

3

incidence of flooding of lands adjacent to converted farms through blocked tributaries as a result of incorrectly sited pond bunds and levees.

When physical habitat changes become disproportionately high through the excessive creation of pond wetlands and the over-concentration of farms within enclosed areas such as lagoons, ecological imbalance occurs and ecological functions are disrupted. These converted habitats are then no longer able to buffer or counteract declining water quality in the same way as natural habitats. Bad farm management practices lead to further reductions in water quality. In Sri Lankan tiger prawn farming, poor water quality led to slower growth and reduced marketable size from 40 to 28 g over several growth cycles. This was followed by disease outbreaks and depressed production, and ultimately to the closure and abandonment of saline prawn ponds, which create further physical change to wetland habitats.

3.2 Mangroves

Mangroves and aquaculture
There are about 15 million hectares of mangrove brackish wetlands in the world, nearly half in the Indo-Pacific region, the rest in Africa and the Americas. Mangroves are increasingly recognized as important reservoirs of biodiversity as well as providing nursery grounds for coastal fish and prawns.

Mangrove wetlands are important nursery areas for fish, and barriers against storm surges

With the development of cash economies and urban existence, mangroves have been under increasing pressure over the last century. The bulk of the damage has stemmed from overuse of wood for fuel, wood pulp and timber, clearing for saltpans or agriculture, coastal development for housing and tourism, plus diversion of freshwater supplies away from the mangrove communities. However, in the past 25 years a new threat has emerged as mangroves have been modified for prawn farming.

Because mangroves have usually been regarded as relatively poor land, with unpleasant living conditions, it has tended to be common land, historically used by subsistence communities with negligible political clout. It has therefore been vulnerable to purchase by outsiders, whether wealthy nationals or international companies. Cheap mangrove land has been purchased repeatedly, cleared for prawn or fishponds and often abandoned after failures due to disease or new competition. Ecuador, Indonesia and Thailand are amongst the most prominent early examples, but this 'gold rush' approach has spread to several other countries (e.g. Sri Lanka in the late 1990s; Figure 8).

More than 250,000 hectares of mangrove, more than half of the total present 80 years ago, have been destroyed in the Philippines, 60% of the loss attributed to the coastal culture of prawns and milkfish (*Chanos chanos*). In Thailand an estimated 25% of the mangrove resource has been lost as a result of aquaculture development. By the mid 1990s Indonesia had already cleared 300,000 hectares of mangroves for prawn farming, but had plans to raise this to 1 million hectares. These clearances have had devastating effects on local fisheries, have marginalized subsistence communities and contributed substantially to coastal erosion, which in turn has smothered coral reefs, creating more ecological degradation. Abandoned prawn ponds are generally of little value because of salination, sulphuration and acidification (see below). In Thailand some 20% have been turned over to salt production and 33% to low intensity prawn culture. The remainder mostly lie idle.

3

Figure 8. Sri Lankan contrasts. A. Lagoon lined with mangroves and inner coconut palms; typical habitat modified for prawn farming. B. Productive prawn farm using good management practices including use of water recycling and probiotics. C. Prawn farm abandoned because of disease resulting from poor management. (S.U.K. Ekaratne.)

Mangrove soils are mostly acid-sulphate soils that exist in a reduced form, so that they are known as 'potential acid-sulphate soils'. When these soils are dug up and ponds constructed, potential acid-sulphate soils become oxidized, and form <u>actual</u> acid-sulphate soils, leading to acid pollution and to destruction of remaining protective mangrove forest outside the area immediately affected. Later on, it is not possible to convert actual acid-sulphate soils back to their original potential acidic status.

Abandoned prawn ponds can also alter physical processes such as sedimentation rates, hydrodynamics including tidal regimes, and nutrient flows. The increased acidity of abandoned pond waters can also affect habitat function, reducing nursery and breeding functions, affecting the feeding and breeding migrations of fish and invertebrates, and changing faunal distributions. Toxic levels of

aluminium may also be released (as occurs with forest acidification in the northern hemisphere). Although the effects of acid-sulphate soils are mitigated by lime addition during prawn culture, this does not take place once ponds are abandoned. Apart from changed water characteristics, mangrove clearance and pond abandonment can lead to other physical impacts. Increased surface run-off accelerates soil erosion, soil organic matter can be increasingly leached and mineralization enhanced. Principal nutrient flows can be altered and soil water storage capacity reduced.

3.3 Artificial habitats

Fish farms
Aquaculture activities often require the establishment of artificial structures to concentrate the target species at the farm location or to facilitate stocking, maintenance,

3

Fish and mollusc farm structures attract fouling organisms, increasing local biodiversity

and harvest. Cages and fences avoid the uncontrolled distribution of fish in vast water bodies. All structures modify the physical environment at the farm location. Light penetration, current speeds, and wave action are reduced by these man-made obstacles. Moreover, artificial surfaces provide substrates for seaweeds and animals, and they increase the physical complexity of the habitat. Artificial illumination may affect daily and seasonal rhythms and predator-prey interactions between the species living near the farm.

In assessing the environmental impact of aquaculture facilities it is often difficult to distinguish between the effects of artificial underwater structures and effects of other aquacultural activities. Artificial reefs, which themselves can be considered as an extensive form of aquaculture, are a more suitable subject for studying the impact of underwater structures. Investigations on artificial reefs, constructed of natural and synthetic materials including tyres and car bodies, have shown that an increased colonisable surface and habitat complexity boosts biodiversity, especially when hard substrata are rare in the natural aquatic environment.

Hard substrata are a prerequisite for the colonisation of a huge variety of seaweeds and sessile animals, such as barnacles, sponges, corals, and moss animals (bryozoans). Like natural substrata, submerged artificial structures become covered by living organisms which themselves increase the complexity of the habitat and attract species of higher trophic levels, grazing and preying on the attached flora and fauna, respectively. Interstices between sessile organisms are used as shelter by mobile invertebrates (e.g. crustaceans, worms, and snails) and by small-sized fish, especially gobies and blennies.

In addition, bigger fish are attracted by artificial constructions. Small-scale fishermen from all over the world take advantage of this behaviour and bring out their gear preferably near wrecks or purpose-built fish aggregating devices (FADs) on the seafloor or floating at the water surface. According to literature reports, the increase in fishing yield near FADs ranges from nothing to 4000%. Whether FADs only attract fish or also improve recruitment and food fish production by providing food and shelter is still under debate.

A typical net-cage for coastal culture of salmon, sea bass and sea bream consists of a square or circular net enclosure covering an area of 25-300 m^2 and penetrating to depths of 5-15 m. The net is attached to a flexible support structure of steel or plastic with buoyancy devices and a walkway. Weights are used to keep the net enclosure in shape. Several cages are grouped and positioned with heavy moorings. Additional nets may be necessary to protect the farmed fish against seals or other predators.

Extensive growth of organisms on the nets (fouling) cannot be tolerated by fish farmers, mainly because clogging leads to oxygen depletion inside the cages. Depending on variety of factors, like mesh size, cultured species, geographical region, and season, nets have to be changed and cleaned at intervals varying from weeks to months. Antifouling agents are used to expand these intervals (see Section 4.3). The organisms attached to support structures, moorings, and buoys may grow undisturbed for years. Marine biologists are currently investigating the cost benefit of whether additional structures should be placed around fish farms to provide extra surfaces for plants and sessile animals. Sessile organisms remove nutrients and suspended particles from the water column, possibly acting as living

3

filters (biofilters) mitigating adverse effects from surplus food, fish excrements and soluble excreta (see Sections 4.1 and 4.2).

Night-lights at fish farms may simply be a measure against vandalism and theft, but there are also biological reasons. Artificial lighting may enhance growth of salmon by expanding feeding times and by stimulating desired hormonal reactions. Mediterranean fish farmers found empirically that, during New Moon nights, artificial illumination of the cages reduced the mortality of sea bass (*Dicentrarchus labrax*). Various species of zooplankton and wild fish are attracted by artificial light and can be used by the cultured fish as supplemental food.

Generally, it is accepted that for ongrowing fish in net cages the advantages of feeding on wild prey are negligible compared to the risk of transfer of diseases and parasites from the wild animals to captive fish. However, attraction of planktonic prey with lights may be of interest, in special forms of aquaculture, e.g. for the rearing of juveniles that depend on zooplankton prey that cannot be cultivated in tanks. The effects of night-lights are not restricted to the cultured fish, as they may also affect activity patterns, orientation, predator-prey interactions, and growth of wild species. There is little evidence that the impact of night lighting has implications on a scale greater than a few hundred metres around fish farms but this issue has not yet received sufficient scientific attention to exclude a potential environmental risk.

Inter-tidal shellfish cultivation
Bivalve molluscs like mussels, oysters and clams are able to close their shell valves to make a watertight seal when out of water. They are filter feeders which means that they can obtain all their nutritional requirements from plankton. These attributes along with their relatively sessile nature make them ideal candidates for intensive farming. Shellfish farmers have developed various techniques for growing them in the inter-tidal zone. One of the oldest coastal farming traditions in Europe is found in France where upright poles called *bouchots* are used to farm the blue mussel, *Mytilus edulis*. These 2-3 metre high poles are arranged in extensive arrays on gently sloping beaches. They are totally exposed at low tide when husbandry and harvesting take place using tractors, which access the crop via regular laneways. This is analogous to terrestrial plant farming and because it is a long established practice (since the 1650s) local stakeholders see it as part of the landscape and an integral part of the local economy. It is a monoculture and the main environmental impacts are competition for primary productivity, siltation from faecal material and domination of both beach space (low-tide) and coastal water (high-tide). In addition the compaction resulting from regular mechanical traffic has a negative impact on burrowing species like molluscs and crustacea. However since sandy beaches are relatively species poor when compared with other marine ecosystems it could be argued that bouchot farming actually increases biodiversity by providing a variety of hard substrata for epifauna and flora to colonise. The poles are in effect vertical mussel reefs with a diverse range of invertebrates and algae, which attach to the mussels and inhabit the interstitial spaces. The seed mussels are usually collected *in situ* by natural settlement, which means that the operation is self-sustaining without need for hatcheries or importation of seed from other regions.

Over 120,000 tonnes per year of Pacific oysters (*Crassotrea gigas*) are produced inter-tidally in France alone, and the same husbandry techniques have been introduced to North

Intensive mussel or oyster culture can remove enough primary production to change pelagic and benthic ecosystems

3

America and other European countries like Ireland and the United Kingdom. The oysters are stocked in plastic mesh bags which are strapped to galvanised iron trestles about 0.5 metres off the seabed. Since these oysters are reef-forming organisms they can be stocked at high densities (10-20 tonnes per hectare) and in areas like Arcachon Bay (France) 7% (10 km^2) of the total inter-tidal area within the bay is occupied by trestles. This has led to local problems of overstocking, which have manifested themselves as poor growth and survival of the oysters and high rates of infestation by parasitic polychaete worms like *Polydora*. In such situations the main environmental issues are similar to the bouchot culture in that when carrying capacity is exceeded there is severe competition for phytoplankton and also a reduction in small zooplankton that is now known to be an integral part of bivalve mollusc diet. In addition there will be increased biodeposition underneath the trestles, which in turn may lead to a 3-4 fold increase in meifauna and a 50% loss of macrofauna when compared with control sites. Because water currents are reduced by the physical presence of the trestles there will also be a local reduction in the depth of the oxygenated layer of sediment. This overstocking has become so serious in France that farmers have been forced, by legislation, to reduce stocking density and IFREMER now monitor the performance of oysters using their REMORA programme.

The other main species grown intertidally are clams like *Ruditapes philipinnarum* (Manilla clam) and the cockle *Cerastoderma edule*. Since these bivalves are naturally found buried in the substratum they do not perform well when kept in suspended culture. Since productivity is generally higher in eutrophic estuaries many oyster and clam farms are located downstream of major river catchments. Here the preferred

husbandry is to plant them in inter-tidal plots and cover them with plastic mesh for the first year in order to reduce bird and crab predation. A low mesh fence, which reduces crab predation and prevents washout of the clams, may demarcate these plots.

Over-wintering flocks of oystercatchers (*Haematopus ostralegus*) can be a problem in Northern Europe and some shellfish farmers have resorted to extreme bird scaring measures like gunfire in order to protect their crop. Another problem arises from the use of the protective netting, which quickly becomes fouled with the green seaweed *Enteromorpha*. This necessitates regular mechanical cleaning and regular net changing which increases compaction of the substratum and disturbance of wading birds. Harvesting is carried out with a mechanical collector that digs up the sand and clams and subsequently sorts and grades them. Various studies have shown that although this activity reduces the density of all burrowing species they will return to normal after 12 months if the site is left fallow.

In some parts of North America both oyster and clams are ongrown in areas of foreshore that have been modified by the addition of gravel and/or shell to stabilise the substratum and give protection against predation. In some areas such as Willapa Bay, Washington, the insecticide carbaryl is sprayed on intertidal oyster grounds to kill burrowing shrimp (*Neotrypaea californiensis* and *Upogebia pugettensis*) which destabilise the sediment. Although this is only carried out every 6 years the combination of chemical control and regular harrowing will almost certainly modify local invertebrate communities.

Both oyster and clam farming often rely upon hatchery seed and so have an indirect impact beyond the immediate environment. Their

3

Too intensive bivalve culture causes poor growth of stock and eutrophication, so there are no winners

ongrowing is a two dimensional activity and so requires large areas of inter-tidal habitat that potentially bring the farmers into competition with coastal birds, tourists and others needing access to estuarine habitats. Very few studies have been carried out on bird–aquaculture interactions. It is obvious that human husbandry activities will disturb birds from time to time, but since inter-tidal shellfish farmers only work their plots on low spring tides this interference is restricted to 4 hours per day for 4 days each fortnight.

Provisional studies in Ireland on the impact of trestles have shown that most bird species are unaffected by the presence of bags on trestles except the dunlin (*Calidris alpina*) which was actually attracted to them! This is one area which needs more research since shellfish farmers are being kept out of protected seabird areas (SPAs) on the precautionary principle when in fact some of their activities might be compatible with avian conservation. For example: local organic enrichment and the provision of hard substrata will increase invertebrate productivity thus providing food and shelter for foraging waders.

Longlines
The world harvest of mussels is over one million tonnes. Approximately two thirds is dredged from the seabed while the remaining third is cultured in suspension on rafts or floating longlines. In Spain about 120,000 tonnes are produced per year from rafts moored in just three Rias (fjord like inlets) on the Atlantic coast. Such a concentration of filter feeding bivalves has a very significant impact on local ecosystems.

A culture of 50,000-60,000 oysters can extract 75-95% of the seston (fine living and non-living particles, predominantly phytoplankton) present in the currents running through it,

indeed the intensively farmed Ria de Arosa of Spain contains so many mussels that they have replaced copepods as the main pelagic grazing organisms. Since copepods are major food sources for fish fry, adult fish and the young of several crustaceans, this implies substantial effects at higher trophic levels, though these effects would be difficult to quantify accurately.

In the last few years it has also been realized that bivalve molluscs inhale (and often ingest) quantities of zooplankton (including copepods, fish eggs, young annelids and bivalve larvae) as well as phytoplankton. Calculations indicate that a single large mussel can inhale as much as 100,000 molluscan and crustacean larvae in a single day, most of which will be killed. The ecological importance of this is as yet unclear, but it seems probable that large-scale mussel/oyster culture reduces recruitment to other benthic invertebrate and fish populations.

The creation of large-scale bivalve farms also has secondary effects on zooplankton. Oyster or mussel culture is characterized by the development of associated fouling communities that also filter the plankton. Barnacles are particularly obvious members of such communities (and take both phytoplankton and zooplankton), but some surprising species thrive in these situations. In the Ria de Arosa and also in the mussel farms of Bantry Bay in Ireland, the small filter-feeding porcelain crab *Pisidia longicornis* (normally an insignificant inhabitant of lower-shore rock pools) occurs in huge numbers among the mussels and their supporting structures.

The mussels also provide a vast surface area which attracts a dense attachment of epifauna. Demersal fish and crabs are attracted to these fouling organisms and also to fallen mussels

3

and detritus on the seabed around the longlines. In extreme circumstances where there is overstocking and poor water exchange anoxic sediment covered with fungus (*Beggiatoa* sp.) can build up under the lines. This gives rise to the low diversity and biomass of a classical 'organic pollution' infaunal benthos community. Since mussels will not thrive under such conditions either there are no winners. If the polluted site is left fallow it will normally recover fully within 18 months.

Nowadays Environmental Impact Assessments (EIAs) are required for all commercial farms and these would include a measurement of local flushing rates which would have to be sufficient to prevent a build up of detritus from the proposed farm. On the plus side mussels feed on phytoplankton and excrete high levels of ammonia which increases the rate of

geochemical cycling of nitrogen. One of the effects of this is to boost primary production in the form of macroalgae growth. It had been estimated that seaweed production associated with mussel rafts (0.5-6.0 g C m^{-2} raft day^{-1}) accounts for about 8% of the total primary production in the Rias.

Mussel longlines and rafts also act as FADs (floating attracting devices) which harbour large populations of fish such as blennies (*Blennius* sp.), bib (*Tricopterus* sp.) and gobies (*Gobius* sp.). These in turn attract fish eating birds such as cormorant (*Phalacrocorax carbo*) and shag (*Phalacrocorax aristotelis*) that find the longline buoyancy barrels very convenient perches for both fishing and preening activities. The importance of mussel rafts and longlines as alternative habitats for fish and birds has not been quantified, but it is obviously significant.

Summary

- Considerable damage to coastal ecosystems, particularly coastal mangroves, has occurred through the development of pond aquaculture, particularly for penaeid prawns.

- As well as direct extensive mangrove clearance, pond culture has resulted in soil salination and irreversible soil acidification and sulphuration.

- Damage to coastal wetlands has caused significant declines in coastal fisheries, and indirectly resulted in the destruction of coral reefs because of smothering following coastal erosion.

- Artificial structures used in aquaculture affect local ecosystems. They provide substrara for fouling organisms such as seaweed, barnacles and mussels, and can function as artificial reefs to attract and aggregate fish.

- Large-scale bivalve farms have secondary effects on marine ecosystems by filtering phytoplankton and zooplankton out of the water column, substantially altering local food webs.

4 The effect of aquaculture wastes on ecological systems

The culture of herbivores globally typically generates few waste problems as naturally available biomass is consumed. The harvest removes carbon and nutrients from the aquatic ecosystem, although there may be some local accumulation of waste products. For example, mussels in suspended culture filter particulate organic material from the water column and shed deposited faeces and pseudofaeces on the seabed (see Section 8.3 for more detail). In the extensive pond culture of omnivorous fish, such as some carp species, fertilisers may be added to the pond to stimulate production on which the fish will feed, but wastes are generally recycled.

On the other hand, the intensive culture of carnivorous species such as tiger prawns and salmonids requires large inputs of external nutrition. The conversion efficiency of feed to aquaculture product is around 20% at best, so 80% of the inputs are wasted. These waste products can be soluble or insoluble (solids). If untreated, the solid (particulate) wastes degrade, adding components to the dissolved pool. In tanks and to some extent in ponds, it is possible to separate some or all of the solid wastes and treat or recycle them. On economic grounds, cage culture is preferred for many species as both solid and dissolved wastes are normally released directly into the environment, there are no waste treatment costs and no expensive pumping of water.

Culture of aquatic herbivores usually creates little waste

Figure 9. Pellets beneath fish cage being fed on by wild fish. (Helmut Thetmeyer.)

4

4.1 Particulate wastes

The major particle effluent from a cage farm consists of faecal material and uneaten fish feed (Figure 9).

The amount of faeces and feed will depend not only on the digestibility of the food, but also on a range of other environmental and husbandry factors such as temperature and disease status. Although trash fish (i.e. by-catch unfit for human consumption) are still used in many developing countries as a source of fish feed, this has been replaced in more developed areas by processed fish feeds, often specifically designed for the culture species. In general these are fish meal/fish oil based, but they also contain a wide range of components including wheat, soya meal, crustacean meal, vitamins, amino acids, minerals, pigments and probiotics.

Such modern diets are easily assimilated and give good feed conversion ratios (FCR: product produced per unit feed). This development, from utilising readily available local feeds to highly developed high-quality feeds, has reduced waste inputs to the environment per unit production. Trash fish rapidly break up during the feeding process with much loss of uneaten particles. They are of variable freshness and nutritional quality, leading to high faecal outputs. To compensate, farmers have developed increasingly sophisticated feeding controls and management strategies that reduce over-feeding.

Economics are also important, as overfeeding is most likely when the value of the product is high and the cost of the feed is low, with greater care being taken of an expensive feed product. In the early years of the Atlantic salmon farming industry, feed losses were up to 20%. It is now generally accepted that feed losses have been reduced to less than 5% in well-run

farms. This is important, as fish feed is extremely energy-rich, causing much greater organic enrichment than faeces on a weight for weight basis.

The solids emanating from cage farms consist of a range of particle sizes and densities, with a range of settling velocities. These particles are affected by water currents that may vary with depth, so can settle well away from the farms themselves, but usually the highest deposition rates are in the immediate vicinity of the farm. The eventual site of deposition will depend on local bathymetry, water movement, and flocculation (clumping of finer particles to form larger, more rapidly settling particles). Bacteria may break down slow settling particles, leading to the release of nutrients into solution. A variety of computer models have been used to track particles to the bed in an effort to predict the zone of organic enrichment. On reaching the seabed, these particles may become incorporated into the sediment or may be resuspended by near-bed currents, thus further dispersing them away from the cages (Figure 10).

Addition of organic wastes to sediments immediately causes an oxygen drain as the wastes are degraded by bacteria. The oxygen concentration at any point in the sediment is dependent on the rate of its uptake, either to fuel aerobic metabolism, or to re-oxidise reduced products released by anaerobic bacteria deeper in the sediment. When the oxygen demand caused by the input of organic matter exceeds the oxygen diffusion rate from overlying waters, sediments become anoxic and anaerobic processes dominate. As sediments become more reducing with increasing distance from the water column/sediment interface, a range of microbiological processes become successively dominant in the following order:

Intensive culture of carnivorous fish or prawns risks organic pollution from uneaten food or faeces

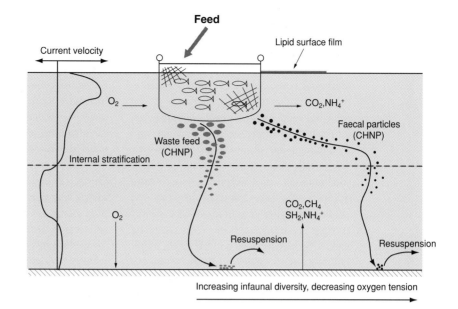

Figure 10. Fate of inputs to and outputs from fish cage culture systems.

- Aerobic respiration, ammonium oxidation to nitrite and nitrite oxidation to nitrate (these aerobic nitrifying processes are inhibited by sulphide and are of limited importance in sediments beneath marine fish farms);

- Denitrification (producing dinitrogen from nitrate);

- Nitrate reduction (producing ammonium from nitrate) and manganese reduction;

- Iron reduction;

- Sulphate reduction (producing toxic hydrogen sulphide);

- Under the most reducing conditions, methanogenesis (producing methane).

In marine systems, sulphate reduction is the most important terminal anaerobic process for organic matter degradation, but is much less important in fresh water due to the low ambient sulphate concentration.

The Redox Potential profile measured down the sediment column (to a depth of 10-15 cm) is widely used to assess the degree of carbon enrichment in the sediments. Positive Redox Potential values indicate aerobic conditions, while negative values are associated with anaerobic microbial processes. Under normal conditions of detritus falling to the seabed, carbon input is moderate and the Redox Potential Discontinuity Level (RPD), i.e. the point at which anaerobic processes become predominant, lies some centimetres below the surface. As carbon inputs increase, the RPD approaches ever closer to the surface as the BOD (Biological Oxygen Demand) within the sediments increases. Eventually, under very high organic inputs, the RPD coincides with the sediment/water interface; under low flow conditions, it may even make the water close to the sediments anoxic.

Communities of animals burrowing in

4

Cage culture effects on the sea bed are localised, and quickly reversed by fallowing

sediments, which receive normal detrital inputs, are species rich, have a relatively low total abundance/species ratio and include a wide range of higher taxa, body sizes and functional types. As organic inputs increase, this diversity also initially increases as the enhanced food supply provides opportunities for the expansion of existing populations and the immigration of new species. However, deterioration of the physical and chemical conditions in the sediments progressively eliminates the larger, deeper-burrowing and longer-lived forms favouring smaller, rapidly growing opportunist species. With increasing inputs, the surface sediments become anoxic and only a small number of specialist taxa can survive, mainly small annelid and nematode worms, which may flourish in huge numbers. Where anaerobic processes occur close to the sediment surface, this may become covered in dense white mats of sulphide oxidising bacteria *Beggiatoa* sp. High flow rates, bringing a continuous supply of oxygen to the sediment surface, do allow the survival of infauna even

when the sedimentary surface layer are anoxic but, where sediments suffer oxygen deficiency for even relatively short periods of a few hours, e.g. caused by slack water, large sections of the benthic macrofauna are eliminated. Ultimately, increasing levels of sedimentary BOD bring about anoxia in the lower levels of the overlying water column leading to the elimination of all metazoan taxa (Figure 11).

Organic degradation rates for labile materials such as are present in waste feed (e.g. lipids and protein) are broadly similar in both anaerobic and aerobic sediments but less labile organic material degrades much more slowly in aerobic sediments. The small worms that dominate enriched sediments significantly enhance the degradation rate of organic materials by mechanisms that are not yet fully understood. Thus if these are excluded by severe sedimentary anoxia the rate of organic breakdown is reduced enhancing organic accumulation, an example of a negative feedback.

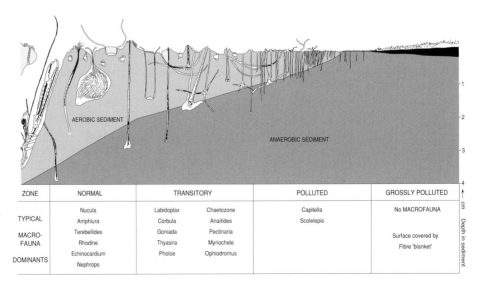

ZONE	NORMAL		TRANSITORY		POLLUTED	GROSSLY POLLUTED
TYPICAL MACRO-FAUNA DOMINANTS	Nucula Amphiura Terebellides Rhodine Echinocardium Nephrops		Labidoplax Corbula Goniada Thyasira Pholoe	Chaetozone Anaitides Pectinaria Myriochele Ophiodromus	Capitella Scolelepis	No MACROFAUNA Surface covered by Fibre 'blanket'

Figure 11. A transect across N.E. Atlantic sediments after organic enrichment. (Courtesy of Tom Pearson.)

4

Seagrass meadows stabilize sediments

Benthic flora, as well as fauna, may be affected by sedimentation from cage farms. Meadows of the seagrass *Posidonia oceanica* cover vast areas in shallow regions of the Mediterranean. They are regarded as the cornerstone of the littoral ecosystem providing a wide variety of niches, accounting for the high diversity of these areas. Seagrass meadows are feeding grounds for green turtles, *Chelonia mydas* and also important nursery grounds for fish and crustaceans. Nutrient inputs from seabass cage farms have been observed to result in an increase in leaf length and increased biomass of epiphytes (algae growing on the seagrasses) and fish, but with decreased meadow density and total disappearance of *Posidonia* directly beneath the cages.

The rate at which sedimentary ecosystems recover following the removal of cages or the cessation of farming is of considerable interest, particularly as the fallowing of sites and rotation of cages has now become recommended practice in many areas. At an intensive farm in the Mediterranean, 11 months after removal of cages, most geochemical variables in the sediment 10 m from the edge of the cage site had values similar to those found at the reference station 1.2 km from the site. However, values at the centre of the cage site were still showing large fluctuations indicative of continuing biogeochemical disturbance even after 23 months. Recovery appears to occur in fits and starts with apparent return to normal followed by a collapse back to impacted conditions often triggered by natural inputs of organic detritus.

In a Scottish study of benthic recovery, communities adjacent to the cages returned to near-normal 21-24 months after farming ceased, but to date no study has looked at recovery processes over a sufficiently long period to be certain about recovery times. Recovery rates are faster in the warmer waters of the Mediterranean and the South China Sea than in the boreal North Atlantic, while local hydrographic conditions also have a considerable influence.

In summary, particulate wastes from cage farms have a profound effect on the benthic environment and recovery may take several years. However, severe effects are generally confined to the local area (a few hundred meters at most) and the total area of seabed used for this purpose is usually insignificant in terms of the total coastal resource.

4.2 Soluble wastes

Most culture of carnivorous fish species takes place in marine cages suspended in more or less open water. Cage structures are relatively cheap compared with equivalent land based structures and, by being immersed in the receiving environment, cage culture avoids the need for expensive pumping of water to supply oxygen to the fish and to remove waste products. This means, however, that dissolved components are released directly into the marine environment in a highly biologically active form. The dissolved products include ammonia, phosphorus and dissolved organic carbon (DOC). The DOC component itself contains fractions rich in nitrogen (DON) and phosphorus (DOP). These waste products have a variety of sources: they may be directly excreted; they may be dissolved from the feed or from faecal particles, or may be released from particles that have been deposited on the seabed around the cages. Lipids released from the diet may form a film on the water surface, and this is often observed around cages after feeding.

4

Prior environmental impact analysis is now routine in most developed-country intensive aquaculture

Harmful algal blooms have been associated with intensive aquaculture but causal links have not been established

The effects that dissolved wastes have on the environment depend on the speed at which these nutrients are diluted before being assimilated by the pelagic ecosystem. Cage structures are often located in areas of partially restricted exchange (e.g. fjords, lagoons) as such locations generally provide shelter from extreme weather, thus protecting staff and equipment. In restricted exchange environments, it is useful to estimate flushing time (i.e. the time taken to exchange all or the major part of the local water volume with new coastal water) in order to assess the risks of significantly increasing the nutrient concentrations in the immediate environment (termed hypernutrification).

Where the flushing time is less than the typical generation time of phytoplankton (a few days), any increase in nutrient concentrations caused by the farm will be unlikely to lead to measurable increases in the local phytoplankton biomass (eutrophication). However, heterotrophic bacterioplankton have much shorter generation times (hours). Around fish farms these organisms have access to highly available organic nutrients, but there has been little research on their response and on any consequences that increased bacterial/protozoan biomass might have on other compartments of the ecosystem.

It is generally assumed that, in contrast to freshwaters, nitrogen (N) is likely to be the nutrient limiting phytoplankton growth in marine waters. This is not always the case (e.g. in nutrient-poor parts of the Eastern Mediterranean) and alterations in the ratio of N to phosphorus (P) might cause changes in the phytoplankton. Phosphate limitation resulting from natural alterations in the N:P ratio has been implicated in toxin production by phytoplankton, particularly dinoflagellates,

but only rarely has it been possible to demonstrate any linkage between the nutrients produced from farming and a biological response, although many such linkages have been claimed.

Beyond looking at purely local enrichment, it is normally not feasible to attribute wider-scale effects to nutrients from farms. However in the brackish waters of the Finnish archipelago in the northern Baltic Sea increasing levels of eutrophication attributable to nutrients from rainbow trout farms resulted in the rapid growth in summer of algal mats on the bottom sediments, which caused bottom water anoxia and strong reductions in local fish and benthic populations. In Israel, fish cage culture in the Red Sea at Eilat is under pressure from environmentalists and the tourist industry because of proximity to coral reefs showing some evidence of degradation although the causes of this degradation have not been clearly linked to fish farming.

Occurrences of nuisance or harmful algal blooms (HABs) are increasing on a global scale and appear to be increasingly prevalent in areas used for intensive mariculture. However, it is not easy to make causal links, especially as there are reports of harmful blooms that do not appear to be associated with pollution (see Info Box 1). It is possible that there is some link between the apparent increase in HABs and climate change. Increased intensive agriculture also results in more use of fertilisers allowing elevated amounts of N and P to reach the coastal seas via freshwater catchments.

Aquaculture itself is one of the industries most affected by HABs, with many incidences of mortalities of cultured fish. HABs can also lead to cultured or fished bivalves being rendered unsafe for human consumption. In 2000-2001

4

Info Box 1: Possible causes of increases in Harmful Algal Blooms (HABs):

Immediate cause	Possible underlying cause(s)
1. A general and widespread increase in the amount and productivity of all phytoplankton, of which potentially harmful algae are a part	(i) nutrient enrichment* (ii) environmental changes associated with climate change; (iii) decrease in predation by zooplankton on pelagic micro-algae
2. A general and widespread change in the floristic composition of phytoplankton resulting in a higher proportion of potentially harmful species	(i) changes in the ratio of nutrients*; (ii) environmental changes associated with climate change; (iii) changes in the relative predation on different types of pelagic algae
3. Increases in the abundance and/or spread of particular, potentially harmful species	(i) changes to environmental conditions favouring these species; (ii) spread of seed-stock through transfers in ship's ballast waters or in farmed shellfish*; (iii) long-term fluctuations in the viability of populations of these species or in their predators or parasites
4. Increase in the amount of synthesis of toxins within potentially harmful species	(i) changes in the ratio of nutrients*; (ii) genetic changes following sexual reproduction during life cycles that may last years; (iii) infection by bacteria or viruses that either make toxins themselves, or make precursors, or alter the algal genome

*Conceivably linked to aquaculture practices

---

large areas of the Scottish and Irish mussel culture and scallop fisheries were so seriously affected by HABS (particularly amnesiac shellfish poisoning, ASP) that their products were unsaleable. This has resulted in tensions between shellfish and finfish farmers, particularly in Scotland where environmentalists and the media have connected the great expansion of salmon farming (with a waste output comparable with that of the human population of Scotland) with the development of regular HABS.

Between 70% and 80% of the nitrogen added to cage systems is lost to the environment. The majority (50-60% of total N) is lost in dissolved form either directly from the fish or by benthic flux from solid waste beneath the cages. Nitrogen (and P) in feeds has decreased as manufactured feeds become better tailored to the dietary requirements of the culture species. For salmon, modern diets tend to contain more lipids and less protein than previously. This has resulted in a general reduction in feed conversion ratios. These

4

Recirculating aquaculture
systems can ameliorate
environmental pollution

currently approach 1:1, although more efficient feeding methods also play a part in this. The net effect is a reduction in N released to the environment per unit production. Exact figures depend on the husbandry regime but around 70 kg N is lost to the environment per tonne of salmon production. This is likely to be much greater for species with higher food conversion ratios or where feeding practices are more wasteful.

In marine salmonid farms, the environmental losses of P are around 20 kg tonne^{-1} fish produced, 30-40% of which is released in dissolved form with the remainder lost by sedimentation. Between 4% and 8% of the sedimented P is returned to the water column per year. In freshwater, P is normally the nutrient element limiting algal growth but, in some systems with either oligotrophic or eutrophic conditions, N may be the limiting nutrient so it is also necessary to consider management of N inputs to freshwater as well as P inputs in some circumstances.

Recently excavated extensive shrimp ponds have acidic effluents (pH 2.7-3.9) so large quantities of lime are added together with fertilisers to stimulate phytoplankton blooms. In semi-intensive or intensive culture this, together with added feed and faecal production, provides highly enriched sediments and nutrient-rich effluent waters that have high concentrations of suspended particulates that may be hypoxic and have a high BOD. These reduce the quality of the receiving environment and often allow transfer of diseases between farms.

Recirculating systems mitigate some or all of the negative environmental aspects associated with wastes from the culture of carnivorous species. They also have the advantage of

the controlled confinement thus minimising gene and pathogen transfer. Such systems potentially allow complete control of the rearing environment, maximising production as well as allowing for the processing and recycling of some or all of the waste products. A key benefit is the minimisation of water use. The disadvantages are in construction and operating costs, but such systems are likely to become more competitive as the environmental costs of waste treatment for all aquaculture types are increased by tighter regulation.

Recirculating systems are built to a wide range of specifications and are undergoing continuous technical improvement but, in essence, these involve a number of key processes including: sterilisation and adjusting both the gas balance and temperature of influent water; maintaining appropriate gas tensions in the culture vessel; removing particulate material by a wide variety of techniques (depending on particle size); and removing dissolved nutrients by bacterially-mediated biological filtration, followed by return of treated waters to the inflow system. Heat can be transferred between system components by means of heat exchangers and collected solids can be stripped of energy by fermentation to provide methane, which can be used as a fuel. The inorganic solid residue can be recycled as agricultural fertiliser.

4.3 Toxic/chemical wastes

Intensive aquaculture is dependent on the use of medicines and chemicals to control the biological environment within the culture system. A wide range of chemicals is used including antibiotics, antiparasitics, fungicides, herbicides and disinfectants. The degree of environmental damage of these compounds will depend on their toxicity to local species,

4

their distribution in the environment and their half-life in that environment. Usually, those organisms most closely related taxonomically to the target organisms for any treatment will be most at risk. For example, chemicals used to kill crustacean parasites such as sea lice are likely to pose most risk to wild crustaceans, but the exact effect will depend on the mode of action of the medicine. The more persistent the chemical is in the environment, the greater the ecological effect.

Although vaccines are now available for some of the more important fish diseases, antibiotic compounds are still widely used in both fish and shrimp culture to treat bacterial infections. Around the world, several major antibiotic drug groups (lactams, aminoglycosides, tetracyclines, macrolides) and synthetic antibacterial agents (sulphonamides, potentiated sulphonamides, nitrofurans and quinolones) are used to control bacterial disease of farmed aquatic animals, mainly fish and penaeid prawns. In descending order of risk of environmental contamination, antibiotics are administered directly into earth ponds, as bath treatments or in compound diets. In-feed administration can lead to losses to the environment in uneaten food and in faeces if not completely absorbed or metabolised. Environmental concerns fall into three categories: disturbance of neighbouring aquatic microbial ecosystems (inadequately studied), development of resistant bacterial strains that threaten aquaculture operations, transfer of resistant strains to the human food chain with potential loss of effective medical antibiotics.

The best-studied antibiotics are those used in temperate aquaculture and of those oxytetracycline and oxolinic have been extensively researched. Only a small fraction of the waste antibiotic budget in marine cage farming can be accounted for in sediments, the sink for the remainder being undetermined. Oxytetracycline and oxolinic acid appear to have rather long residence times in sediments, with various half-lives reported up to hundreds of days. In contrast, furazolidone and amoxicillin have only short environmental half-lives (hours). Of the potentiated sulphonamides, sulphadimethoxine and ormetoprim appear to have short half-lives, as does trimethoprim, but sulphadiazine is relatively persistent. Care must be taken when comparing half-lives of antibiotics as the statistic may refer either to the chemical stability of the compound or to how quickly it is dissolved out of sediments into pore water and lost to the overlying water column.

Anti-bacterials may be found in non-target organisms around farms but the amounts found vary between species, presumably due to their different feeding habits. After antibiotic treatment at an Atlantic salmon site in Puget Sound, only trace levels of oxytetracycline were found in Dungeness crabs and oysters from beneath the cages, but high levels in red rock crab. The occurrence of antibiotics, and bacterial resistance, has been observed in wild fish and mussels near farms after treatment, but typically antibiotics are only found in animals in close proximity to the farm. Disturbingly, a quinolone antimicrobial agent has been found in saithe (*Pollachius virens*) outside salmon pens in Norway at levels well above those allowed for human consumption in that country.

In a study of antibiotic residues and bacterial resistance in sediments beneath a marine cage salmon farm, oxytetracycline was confined to an area of the sediment that was smaller in extent than the area of the cage block itself. Elevated frequencies of resistance were detected in samples from a wider area than the cage block, but there was no correlation between the

4

concentration of oxytetracycline in a sample and the frequency of resistance that was determined in the culturable microflora present.

Antibiotics can affect sedimentary biogeochemical processes, presumably by interference with bacterial ecology. Seven days after application of oxytetracycline, oxolinic acid, or flumequine to sediments in experimental tanks, sulphate reduction rates were reduced to less than 10% of the rates found in controls. This effect lasted at least 29 days but was absent after 70 days. In a similar experiment with potentiated sulphonamide (trimethoprim and sulphadiazine), amoxycillin, oxytetracycline and oxolinic acid, the potentiated sulphonamide was found to have the greatest effect on sediment aerobic respiration rate and amoxycillin the least. In addition, oxytetracycline has been shown to inhibit bacterial nitrification, which could lead to the build up of ammonia and nitrite in sediments.

In the early 1990s, antibiotic residues in cultured shrimps were commonly reported and shrimp consignments from Thailand and Indonesia have been rejected by Japan because of these. Since the mid 1990s, shrimp exporting countries in Southeast Asia are more aware of problems caused by antibiotic tainting and most of these countries have initiated monitoring programmes to ensure compliance with food safety regulations from importing countries. All antibiotic residues are of concern but some more than others, for example the use of nitrofurans (e.g. furazolidone) is banned in some but not all countries as a consequence of their carcinogenic or mutagenic properties. Shrimp farming is beset by resistance problems. Diseases caused by *Vibrio* bacteria are major problem. Multiple drug resistance has been detected in several strains of *Vibrio*, and transfer of resistance to other types of bacteria

(including *Escheria coli*) via plasmids has been demonstrated.

Some creative alternatives to antibiotic use are being investigated. For example, in Thailand plant extracts are investigated for antibacterial and antiviral qualities – a promising substance is extract of guava (*Psidium guajava*) that appears to prevent bacterial disease in catfish (*Clarias macrocephalus*). More exciting, though so far of limited application, is the probiotic concept. A probiotic is a single or mixed culture of live, naturally occurring microorganisms that have a positive effect when introduced into tanks or ponds. Primarily these act by enhancing degradation of organic matter, improving water quality and releasing antibacterial and antiviral compounds. Some commercially available probiotics have been shown to be effective against *Vibrio* sp., *Aeromonas* sp. and White Spot virus.

The use of large amounts of chemicals in an aquaculture enterprise is usually a sign of crisis and poor husbandry. Where such operations discharge significant amounts of hazardous chemicals to the aquatic environment, it is likely that the operation is unstable and probably unsustainable. Stressed animals are more likely to succumb to disease. However, improved understanding of the environmental requirements of the cultured species, together with immunological developments, e.g. vaccines, can significantly reduce chemical usage. For example, environmentally unsound shrimp farming practices, requiring high chemical inputs, led to massive premature abandonment of intensive culture facilities in Indonesia, the Phillipines and Taiwan. These were not experienced in Thailand where there had been adoption of more environmentally friendly, locally adapted methods involving

4

Figure 12. Sea lice on belly of salmonid fish. (David Hay.)

Sea lice densities in the wild may
be locally increased by intensive
salmonid culture

reduced water exchange and reduced chemical usage.

Sea lice treatments

Sea lice are mobile ectoparasitic copepods, which live on the gills and other body surfaces of fish (Figure 12). They feed on mucus, skin and blood, causing open wounds that expose fish to osmotic and respiratory stress as well as providing a route for secondary infections by bacteria or viruses.

In the sea-cage rearing of salmon, two sea lice species can cause severe infestations, heavy mortality and reduced marketability. *Lepeophtheirus salmonis* is specific to salmonids (Pacific as well as Atlantic) and *Caligus elongatus* is found on over 70 fish species. Sea lice have free-living nauplius and copepodid stages, so it is possible for sea lice from one farm to infect other farms, and for there to be interchange between captive and wild fish. Sea lice therefore present a major commercial problem. For example, the cost to Norwegian and Scottish salmon farms alone amounted to

70 million euros in 1996-1997. Management at farm level is needed in addition to control by regulatory bodies. There is also an environmental concern that declines in wild sea trout and Atlantic salmon populations might be related to parasite transmission from fish farms. Although no conclusive proof of a causal link has been reported, there is widespread acceptance that sea lice from farms affect wild populations. A particular problem is that many farms are sited in those fjords that are the natural migration routes of wild salmonids, which have to run the gauntlet of waters with high levels of sea lice larvae.

Treatment of sea lice has traditionally been by chemical means; this can be done by bathing fish in dilute solutions or, more recently, by incorporating effective agents into pelleted fish diets. Either approach inevitably results in large losses of the medicines to the environment around fish farms. Residues of chemicals in salmon tissues are also a consumer issue. Initially the agent of choice for bathing fish was dichlorvos, an organophosphate. It kills

4

Biological control of sea lice by 'cleaner' wrasse provides an alternative to chemical treatment

adult lice by blocking acetylcholinesterase production, leading to continuous neuromuscular activity and terminal exhaustion. There have been concerns about the release of dichlorvos into the environment, and health concerns for farm workers. Because of this, and also because lice resistance to dichlorvos has been reported since the early 1990s, this product has largely been phased out. A common replacement organophosphate is azamethiphos, widely used in North America and Norway.

A more environmentally friendly bathing chemical is hydrogen peroxide, a common treatment method (at 1500 ppm) in Scotland for several years. While reasonably effective against adult lice, it is lethal to salmon at temperatures above 14°C, and resistance has recently been confirmed for farms with a long history of its use. Hydrogen peroxide has now been almost completely replaced by the pyrethroid cypermethrin, which acts by prolonging normally transient increases in sodium permeability of nerve membranes. Again, it is administered as a bath treatment with the unused chemical being released to the environment. It is highly toxic to crustaceans and its long-term ecological effects are still being studied.

Treatments administered with the diet (in-feed) have the potential advantage of use of much lower quantities of potentially toxic agents. The most well known is ivermectin, one of a family of nerve poisons, the avermectins. Widely used in agriculture by veterinarians to control ticks, initial fish farm usage was informal, as the compound was not licensed for use in fish. Interest in use is strong and ivermectin has been extensively used in Ireland. However, there have been environmental worries and fears of conflict with fisheries since avermectins are highly toxic to crustaceans

such as shrimp and lobsters. Experimental evidence also shows adverse effects on benthic polychaetes. Adverse publicity of illicit use has led to strong consumer resistance via retailers. Ivermectin use in now illegal in the EU but another semi-synthetic avermectin, emamectin benzoate is now in widespread use. Unlike Ivermectin, this product is licensed for use in fish. Similar problems accompany use of chitin-synthesis inhibitors such as diflubenzuron and teflubenzuron that prevent proper exoskeleton formation in arthropods in general.

Considerable research effort has been expended in developing an effective vaccine for sea lice but with little prospect of early success. There are, however, other non-chemical approaches to sea lice control. The simplest is that of fallowing, here a farm site raises one group of salmon smolts to marketable size, then waits 2-3 months before restocking the cages. This results in lower sea lice densities, but reduces farming intensity and is ineffective in areas where there is multiple ownership of nearby sites unless water-body management agreements are in place and practiced. Where management strategies have been implemented, considerable reductions in infestation have been recorded, especially during the first sea year. Another approach is to manipulate feeding regimes to induce salmon in holding pens to spend more time at depth, since most louse settlement onto fish appears to take place in the top 4 m of the water column. So far this technique appears to be a useful supplementary method rather than being effective on its own.

In the late 1980s a Norwegian fish biologist Åsmund Bjordal carried out field trials with salmon using 'cleaner' fish rather than chemicals to control sea lice. A number of fish obtain significant amounts of their diet by

4

Figure 13. Goldsinny wrasse *Ctenolabrus rupestris.* Used to 'clean' sea lice from salmon.

removing surface material (fouling organisms, loose dead tissue, ectoparasites) from other animals, such as fish or turtles. In northern Europe, the main groups of fish that do this are members of the wrasses (Labridae) and several species will remove sea lice from salmon. The idea of biological control was an attractive one and in the last decade, use of cleaner wrasse, particularly the goldsinny wrasse (*Ctenolabrus rupestris*) (Figure 13) has become widespread, especially in Norway and Iceland where it is commonly the primary method of louse control.

At present Norway alone uses around 5 million wrasse per year and fishing for wrasse is highly profitable. However, there are potential ecological problems with the use of cleaner fish. Firstly, the wrasse are collected from the wild and heavy mortalities in salmon pens, particularly in winter, mean that wrasse have to be continually replaced. This raises the possibility of overfishing of wrasse, though this is offset by consumers (i.e. salmon farmers) requiring fish above 10 cm in length (smaller ones would escape easily from salmon pens). Secondly there are concerns that use of wild cleaner fish risks transferring disease, especially Infectious salmon anaemia, to farmed salmon – this risk has been minimized by using only wrasse caught in the neighbourhood of farms. Wrasse culture has been conducted experimentally and the long-term future of biological louse control is likely to lie in the production of cultured wrasse stock that is guaranteed disease-free.

Antifoulants
Except in the case of simple pond aquaculture operations, almost all fish and invertebrate farms around the world involve placing man-made objects (nets, ropes, cages, buoys etc) in water. Fouling of these objects is a major problem for aquaculture, particularly mariculture. Without protection, artificial materials placed in sea water become fouled by bacterial films within hours and the films are enriched by diatoms within days. Over succeeding weeks seaweeds and a variety of animals such as hydroids, barnacles and

4

mussels join these. Fouling organisms clog nets and add weight to structures, making them more vulnerable to bad weather.

Generally fouling takes place more quickly in tropical than temperate areas, but antifouling is a major expense for all intensive mariculture operations. Mostly, fouling is removed manually, by high-pressure washing or steam cleaning. However, net and cage treatments with various toxic chemicals have been used. In the case of culture of bivalve and gastropod molluscs, fouling of the cultured organisms themselves can be a substantial problem – mussels fouled with *Pomatoceros* tubeworms can be unsaleable. In Canada, mussels are sometimes treated with quicklime to remove foulers, probably creating minor local pollution on disposal.

No antifouling technique is effective against all foulers, nor is any antifouling material effective indefinitely. However, in the 1980s coatings based upon tributyl tin (TBT), with some herbicide addition came close to this ideal recipe. Unfortunately TBT has dramatic effects on molluscs at extremely low concentrations (<1 ppb). Female dogwhelks (*Nucella lapillus*) developing penises (the 'imposed sex' or imposex response) have drawn most media attention, but the first effects were seen in cultured oysters which had stunted, thickened shells and poor meat to shell ratios. Although most problems of TBT and aquaculture have revolved around contamination of farms because of nearby harbours and marinas, there have been cases of mussel farms being adversely affected by TBT treatment used at upstream salmon farms. Some TBT residues were reported from salmon flesh many years ago, but TBT treatments have been replaced in aquaculture by much less toxic treatments containing various percentages of copper. The ecotoxicology of copper effluents from cage farms is, however, currently under investigation.

4

Summary

- The intensive culture of carnivorous species generally causes greater environmental impacts than more extensive methods employed in the culture of herbivorous species.

- Carnivorous species are typically fed on food rich in fishmeal and fishoil but also including a variety of vitamins, pigments and trace elements. These energy rich feeds can cause a deterioration in the environment when lost from fish cages and when nutrients and organics are exchanged from pond systems.

- In cage farms in marine waters, waste food and faeces settle on the seabed stimulating microbial activity with a consequent increase in sedimentary oxygen demand and increase in sulphate reduction. This causes major changes in animal populations burrowing in sediment, which become dominated by small opportunist polychaete and nematode species that can be present in huge abundances. Such polluted sediments can be covered in mats of white sulphide oxidising bacteria.

- The degree of impact on sediments and the consequences for benthos are crucially dependent on the rate of input of organic material from the farm and the degree to which water currents disperse and resuspend sediments and the supply of oxygen to meet benthic demand.

- Although effects on sediments can be locally severe, even leading to the total loss of metazoan life, they are generally localised around the farm. The total coastal seabed area polluted by cage fish farming is insignificant in global terms.

- Nutrients are discharged from intensive culture both directly from the cultured species as excretory products e.g. ammonia and urea and from degradation of particulate wastes in the water column or from sediments. Nitrogen is often the nutrient that is limiting for plant growth in marine systems whereas phosphorus limits growth in fresh water. There are exceptions to this e.g. in the phosphorus-limited, nutrient-poor eastern Mediterranean. Modern diets are formulated to reduce P and N wastage.

- The effects nutrients have on the ecosystem will depend on their concentration (which depends on the input rate and the dilution rate for a particular environment) and on what is limiting plant growth (e.g. light, macro or macronutrient and grazing pressure).

- Medicines (e.g. antibiotics and antiparasitics) used in aquaculture can have impacts on ecological processes (e.g. microbial metabolism and crustacean larval development).

4

Summary - (continued)

- Vaccines have reduced the mortality of several serious pathogens to farmed stock and reduced reliance on medicines. Biological controls e.g. cleaner fish for sea lice have been used to reduce dependence on chemical treatments.

- Seal lice are parasites of farmed fish. Salmon lice from farms have been implicated in the declines of wild populations of sea trout and Atlantic salmon.

- Antifoulants can have wide ranging ecological effects. TBT is known to affect marine molluscs at low concentrations and is now banned for use in fish farms.

- Most environmental problems from aquaculture can be mitigated with present technologies. These can range from multi-trophic polyculture systems that maximise the utilisation of inputs, to simple settlement ponds to remove particulate effluents, to near total recirculating systems where wastes are treated and water recycled internally.

5   Diseases and parasites

5.1 Disease as a part of life

Parasites and disease are part of the natural biology and functioning of ecosystems. Disease results from the responses of the body of an animal or human to injury, caused by pathogens, and parasites. Pathogens and parasites are organisms that live at the expense of another organism, the host, and divert the energy of the host away from its growth and reproduction and into resisting the disease. The host resists through its immune system. If the host immune reactions overcome the pathogen, the host returns to normal, to homeostasis. If however the pathogen overcomes the host resistance, disease results, and the host may die.

So whether a parasite or a pathogen causes disease or not, depends on the host's immunocompetence. For example, furunculosis, a disease caused by a bacterial infection, is often found in recently spawned wild salmon. But furunculosis is very rarely seen in other wild salmon, not because they have not been exposed to the bacterium, but because their immune reactions overcome the bacterial infection. Salmon which have spawned on the other hand, have little immunocompetence: they have completed the long journey into freshwaters and upstream to the spawning ground without feeding, and all their energy has gone into the stresses of migration and spawning.

In aquaculture, fish and shellfish are removed from their natural niches, and while the total set of conditions under which they are cultured replicates, as far as possible, the conditions in the wild, this is limited by practical and commercial constraints. Any change in living conditions may cause stress and predispose an animal to disease. Disease requires not just a suitable host and a pathogen, but also a stressful environment to unbalance the host/pathogen relationship, and the host immune system to be overcome at least temporarily by a pathogenic factor (Figure 14).

Environmental stress has several possible causes. For example, biological causes such as overcrowding of fish in fish cages; or chemical causes such as excessive ammonia or other toxic pollutant in the water. There are also physical causes, for example an excessive rise in water temperature; or poor procedures, involving handling of fish, grading, or disease treatment.

In aquaculture facilities therefore, disease can be a significant factor. This creates a situation where pathogens, diseases, or the medications used to treat them, can potentially impact on the environment.

5.2 Examples of diseases and parasites

As aquaculture has been developed worldwide, the species cultured, the geographic spread of

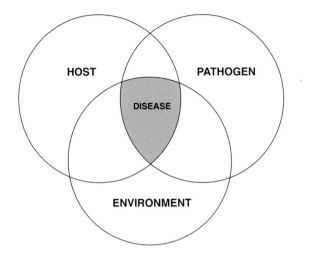

Figure 14. The relationship between host, pathogen and environment and the occurrence of disease (after Snieszko).

the industry, and the variety of environmental conditions in which culture is pursued, have all diversified at an increasing pace, particularly in the past 50-60 years. The species most widely cultivated cover a range of taxonomic groups including oysters, mussels, clams, marine prawns, freshwater prawns, salmonid fish, cyprinids, tilapias, catfishes, snakeheads, milkfish, eels, seabass, seabream, groupers and redfish. The diseases of one group are usually different from those of other groups, and therefore need to be considered species by species. However in general, the causes include microparasites (bacteria, fungi, viruses, protistan parasites), macroparasites such as ectoparasitic lice or endoparasitic nematode worms, and non-infectious agents, such as dietary (neoplasia), or environmental chemical agents, which may cause pathological symptoms including cancer.

The diseases under consideration here are diseases of fish and shellfish themselves. They do not affect humans, even if the fish and shellfish are handled or eaten.

Molluscs

The majority of significant oyster diseases that exist today in oyster growing regions throughout the world have resulted from introductions of parasites related to the movement of shellfish. *Marteilia refringens,* a protistan parasite found in the digestive gland of the European flat oyster *Ostrea edulis,* is the agent causing 'Aber' or 'digestive gland' disease. Since 1968 this parasite has caused serious recurring mortalities in the European flat oyster industry. It was initially diagnosed in oysters in Brittany in France, but prior to this, oysters had been moved to other areas of France and to Spain, where the parasite was also subsequently diagnosed. This parasite remains a problem in oyster stocks in France and Spain.

A second protistan, *Bonamia ostreae,* which infects the blood cells of the flat oyster, was introduced into European flat oyster stocks in the late 1970s. It is thought that the initial introduction of this parasite occurred in a consignment of spat sent from California to

Several diseases and parasites have been spread to wild populations by movements of shellfish

5

France. The disease quickly spread with movements of oysters within France and then throughout Europe, including the major oyster producing areas in France, Spain, Netherlands, England and Ireland, which are still affected today.

Losses attributed to these two diseases continue to have a serious impact on the European oyster industry: for example French annual production of this species has dropped from 20,000 metric tonnes in the early 1960s to approximately 2,000 tonnes today.

In the United States the eastern or American oyster *Crassostrea virginica* has been affected by a number of serious pathogens since the 1950s, predominantly along the east coast. The spread of the two main protistan parasites, *Perkinsus marinus* and *Haplosporidium nelsoni*, has been predominantly related to interstate transplantations of oysters. The extent of the losses caused by these pathogens can be demonstrated by the fact that only 5% of traditional public oyster beds in Virginia are now productive.

Epizootics of a protistan parasite QPX (Quahog Parasite Unknown) have recently occurred in localised populations of cultured hard clams, *Mercenaria mercenaria,* in Prince Edward Island, Canada and Massachusetts and Virginia, USA. This has raised concerns within this industry along the eastern seaboard of the United States, as interstate transplantations of clams are common, increasing the potential risk of disease transfer.

The Pacific oyster *Crassostrea gigas,* which originated in the Northwest Pacific, is rapidly becoming the main oyster species cultured around the world. The literature contains at least three examples of the Pacific oyster serving as a host for pathogens that are similar to, or identified as, agents of oyster disease. For example Pacific oysters in Japan and Korea may have been the source of *H. nelsoni* via unauthorised importations of Pacific oysters to the United States during the 1950s. Introductions of half-grown Pacific oysters into Ireland began in the early 1990s following the free movement of trade within the European Union. Samples of consignments revealed the presence of a number of exotic pest species including *Mytilicola orientalis*, *Mytilicola ostreae* and *Crepidula fornicata*, all capable of becoming established in native populations (see also Section 7.1). There were also large numbers of flat oysters *Ostrea edulis,* a potential vector of *Marteilia refringens,* which is not present in Ireland to date.

Mortalities of over 90% were observed among hatchery reared larvae of Pacific oysters in France and New Zealand initially during the summer of 1991. A herpes-like virus has been found associated with these losses. Elevated water temperatures appeared to favour the spread of the infection, or to accentuate the infection when the animals were already stressed. Concerns were raised due to the large quantities of spat and half-grown oysters transferred within the European Union. However, recent literature indicates that this herpes-like virus has also been associated with mortalities in a number of other species including the native clam *Ruditapes decussatus*, the manila clam *Ruditapes philippinarum* and the mussel *Mytilus edulis.*

Overall, the history of molluscan culture has repeatedly demonstrated that diseases and exotic pest species have been transported around the world, often to the detriment of aquaculture, but also to natural molluscan populations, hence altering ecosystems.

Transport of oysters has often resulted in the transfer of pest species too ('hitchhikers')

5

Crustacea

Crayfish plague has occurred in many parts of the world and is the most serious disease affecting crayfish. It is caused by the fungal microparasite *Aphanomyces astaci*. It is believed to have been spread from North America, where the native crayfish are resistant, but can act as carriers of the fungus. These crayfish have been imported to other countries because of superior growth characteristics. After its original introduction around 1860 to Italy, the plague spread north to both France and Germany and has since been observed in a number of other countries. In Sweden the pathogen has been known to cause 100% mortality, completely wiping out native crayfish from freshwater ecosystems. Since crayfish are an important part of the diet of otters and fish such as the trout, pike, chub and perch, the fishing industry has suffered and lakes have become weedy, and turbid with reduced diversity.

Prawn culture (overwhelmingly of penaeid prawns in marine systems, and *Macrobrachium rosenbergii* in freshwater) has been expanding exponentially in the last two decades. Rearing systems tended to have been overstocked with very high prawn densities, leading to poor environmental conditions, including low oxygen and pH levels, that reduce disease resistance (see Figure 14). In consequence, the growth of the industry has experienced serious checks caused by disease epizootics. Infectious hypodermal and haematopoietic necrosis virus (HHNV) is a parvovirus infecting penaeid prawns on the Atlantic and Pacific coasts, in Asia and the Middle East. It has a wide host range, but is particularly virulent for the larvae and juveniles of *Penaeus monodon*, *P. vannamei*, and *P. stylirostris*. Its geographic distribution has been widening over recent years.

Yellowhead disease (YHD) is the most destructive viral infection in the giant tiger prawn *P. monodon*, causing mass mortality in cultured prawns at grow-out stage. The first epizootic was recorded in Thailand in 1980. Yet another viral disease causing mass mortality not only in the tiger prawn but also in other prawn species, including *P. japonicus*, *P penicillatus*, *P. cinensis*, *P. merguensis,* and *P. indicus*, is white spot disease (WSD). This disease was initially observed in Taiwan in 1991 and spread to China, Japan, Korea and India in 1992. By 1994 it was observed in Thailand and is now present in most prawn farming countries of Asia. Losses due to the disease were estimated at $US 400 million in China in 1993 alone.

By the mid 1990s the disease had spread to the Americas with similar devastating losses being observed. The most likely route for the spread of WSD has been the international trade in post-larvae for on-growing, live brood prawns for hatcheries and dead prawns for processing.

Fish

Diseases affecting fish have continued to emerge, as each new fish species is brought into culture. It must be emphasised that no new infectious or parasitic disease has emerged as a result of aquaculture; each has its origin in wild species. In salmon culture, the viral diseases of infectious pancreatic necrosis (IPN), infectious haemopoeitic necrosis (IHN) and viral haemorrhagic septicaemia (VHS) are long-standing problems; likewise bacterial diseases including bacterial kidney disease (BKD) and vibriosis, or parasitic disease such as whirling disease, caused by the protozoan, *Myxobolus cerebralis*. Spring viraemia (SVC) has been a feature of carp culture.

As culture methods have been developed for an

Prawn culture has been plagued by disease transfer. This has resulted in widespread abandonment of farms

No new disease has emerged because of aquaculture. It is intensive culture that enhances disease transmission

5

ever-wider range of fish species, the range of diseases of concern has grown in parallel. For example, catfish farming in the US is clinically and economically affected by channel catfish virus disease (CCVD), The causative herpes virus affects juveniles, causing high mortality due to loss of osmotic balance. Survivors become asymptomatic carriers. Another virulent problem causing high mortality, which emerged with the catfish industry, is enteric septicaemia (*Edwardsiellosis*) caused by *Edwardsiella ictalurus*. Cultured seabass, turbot and groupers have been seriously affected by the nodavirus infection, viral nervous necrosis (VNN), which can cause 100% mortality in juveniles.

As fish farming has expanded geographically, fish species have been exposed to new pathogens; e.g. piscirickettsiosis is a septicaemic condition of salmon, caused by a rickettsia, and resulting in very high mortalities. It was first described in coho salmon in Chile, but has since been recorded in a range of other salmonids and in other countries. One of the most serious ectoparasitic problems in marine farming is that caused by sea-lice. These copepods are to be found on all marine fish species, but usually in very low numbers. In farms the infestations can become devastating, and the fish tissues are literally grazed away by the lice, causing damage, loss of osmoregulation capability and death. *Lepeophtheirus salmonis* is found only on salmon, but *Caligus* species have a wide host range (see Section 4 for details of anti-lice treatments).

Some diseases result from the culture methods or from diet. For example, *Ichyophonus* causes a fungal infection, which penetrates and destroys the flesh of farmed fish that have been fed a wet diet, derived from fungal-infected marine trash fish. The disease renders the farmed fish unsaleable.

Immune mechanisms

In all living organisms it is the role of the immune system to prevent or overcome pathogenic infections and to maintain homeostasis. Knowledge of the immune mechanisms of shellfish is poor. For more than a century it has been recognised that phagocyte cells play a very important role not only in nutrition but also in defence (see Info Box 2). Phagocytic cells in the body neutralise and eliminate all foreign materials including pathogenic organisms. More recently a large variety of circulating molecules have been recognised in the body fluids, including enzymes, cytolysins, antimicrobial peptides, lectins etc, all of which work to destroy pathogens, and restore health.

In fish, superimposed on the innate defence mechanisms found also in shellfish, is an acquired immune system, which shows the characteristics of specificity, memory and classical immunoglobulin molecules. This is lacking in shellfish. The specific immune response, mediated by lymphocyte cells, equips fish to respond specifically to and target pathogens, and to destroy them rapidly. The fish also retains a memory of each pathogen encountered, so that, should it encounter the same pathogen again, the immune response is faster, greater and more effective.

The existence of a specific immune response in fish has implications for disease prevention in cultured fish because, unlike in invertebrates, vaccination, which is an application of the specific immune response, is a possibility.

5.3 Disease prevention strategies

It is important for the success of aquaculture enterprises that disease is prevented or minimised, because disease can cause loss of productivity, whether through growth,

5

Info Box 2: Phagocytosis:

Phagocytosis is a major line of defence against invading microorganisms, such as bacteria and viruses, and other foreign material. Phagocytic cells are white cells found in the blood and throughout the body. These blood cells are able to differentiate the animal's own cells from foreign cells or parasites within the body, and therefore do not destroy themselves. Phagocytic cells attach to microorganisms or other 'non-self' particles, take them into themselves and kill them. The foreign particle is internalised within the phagocytic cell in a compartment called a phagosome. The phagocytic cell can produce toxic chemicals, including reactive oxygen and nitric oxide, which kill the bacteria and viruses; they also produce enzymes – substances that are capable of digesting away the foreign cells. The digested foreign particles are used by the phagocytic cell as an energy source, and undigested components are extruded through the cell membrane. A thorough understanding of this process has practical applications for disease control in aquaculture.

condition, reproduction or mortality. Disease prevention and control begins with siting farms in locations where the environmental requirements of clean water, adequate flows and exchanges year-round and high oxygen levels at all times can be ensured. Only certified disease-free stock is used; and the genetic strain used is that which thrives best under culture conditions. Diet formulation and processing have to ensure complete nutrition, so that deficiency diseases do not occur. Aquaculture facilities are managed with optimal environmental conditions for the species involved, to minimise stress that in turn can result in disease losses. Thus the interests of the aquaculturist, as well as of the wider community, are best served by maintaining environmental quality in the vicinity of the farm.

Diseases can be prevented using an understanding of the pathogens and parasites, and managing the aquaculture facilities so that the pathogen cannot function. A classical example of this was the elimination of the whirling disease problem, which caused significant losses of young salmonids. Whirling disease is caused by a sporozoan parasite *Myxobolus cerebralis*. The life cycle of this parasite involves development in an oligochaete worm. In the early culture of young salmonids earthen ponds were commonly used, and these ponds had their own flora and fauna, including oligochaetes in the mud. Once the fish culture moved to concrete or other types of construction the life cycle of *Myxobolus* was broken, and whirling disease in culture facilities became part of history.

There are a number of diseases, which pose serious problems for aquaculture, where lack of knowledge of the biology of the pathogen restricts the ways in which the diseases may be controlled or prevented, for example bonamiasis or marteiliosis in oysters, or proliferative kidney disease in salmonids.

5

Diseases can be prevented by vaccination, and as mentioned above a number of successful vaccines are commercially available for some of the more important fish diseases, such as vibriosis in marine-farmed species. Another developing approach, which also exploits knowledge of the immune mechanisms of the animals being cultured, is that of immunostimulation. Immunostimulation involves enhancement of the immune response by administration of harmless compounds such as glucan or laminarin, which increase the activity of some immune functions, and induce more effective resistance to pathogens and parasites.

Once disease occurs, rapid diagnostic methods are required so that appropriate action can be taken to limit losses and prevent spread. In enclosed environments such as hatcheries, culture tanks or ponds, antibiotics can be used against bacterial infections, but they are only used as a last resort and with controlled protocols and prescribed withdrawal periods, to ensure that residues do not remain in the fish flesh, and to minimise the development of resistant strains of bacteria.

Genetic selection of strains of fish and shellfish, which are more resistant to disease, has had considerable success. The further possibility of molecular manipulation of genetic strains, for example by transfer of resistance genes, is not being pursued because of economic forces; in many countries consumers will not buy genetically modified foods at present.

One of the most effective, though devastating approaches to disease control has been the total eradication of affected stock, in a farm where the disease involved is serious, and could be spread by movement of diseased animals or contaminated products or materials to other farms or to other countries. Eradication was used in recent years in accordance with EU Regulations, when Infectious Salmon Anaemia (ISA) was detected in a number of Scottish salmon farms. Movement of fish was stopped; the stock on the affected farms was killed and disposed of, in order that other Scottish and European farms were protected from the possible spread of the disease.

5.4 Impacts of the disease and parasites

There is a clear relationship between environmental conditions and quality, and health and disease in cultured species. Conversely, there are two main areas where disease aspects of aquaculture can potentially impact on the environment: through movement of organisms, and through use of chemicals in disease prevention or therapy. These impacts can arise in situations where aquaculture facilities are managed badly, or where the regulations, controlling the aquaculture industry, are not adhered to or implemented.

The aquaculture industry has depended for its development and diversification on an understanding of the biology of each cultured species, and its subsequent application and adaptation from wild to aquaculture situations. As a result, early fish farming depended on species like carp and trout. Research into newer species for the industry takes time, and as new species have been developed in one country, the tendency has been to export them to other producer countries. Also with specialisation of the industry, some producers have concentrated on the production of spat or juveniles to sell to ongrowers, and again this often involves export of the animals from one country to another. Such movements have in

5

the past involved the movement of pathogens or parasites of a species from a country where it is endemic to a country where it did not previously occur.

For example, the three common viral diseases of salmon culture, are so widespread because of movements of young salmon from one country to another before it was recognised what was happening, and before regulations to prevent such movement had been put in place. Similarly, oyster diseases have been spread by movements of spat (see above). Such movements have serious implications, because in nature host parasite relationships normally evolve over long periods, and in a balanced manner, so that both host and pathogen species survive.

Where a host is exposed to a 'new' pathogen, it is extremely susceptible, because it has no resistance to it, e.g. *Bonamia ostreae* in the oyster *Ostrea edulis*. Furthermore, once a pathogen or parasite has established itself in a new area, it is extremely difficult to eradicate. *Gyrodactylus salaris,* an ectoparasitic worm parasite, is believed to have wiped out many populations of wild Norwegian salmon in the 1970s, when it was transferred from Baltic salmon, which are resistant to the parasite. In recent years whirling disease has become a major cause of loss in wild salmonid stocks in some US rivers, apparently as a result of restocking with hatchery-reared infected stock.

A range of chemicals has been used in aquaculture; for example surfactants and disinfectants to clean facilities, and antibiotics to treat bacterial infections (see Section 4). These are used with care, not only in the interests of the aquaculturist whose fish or shellfish are vulnerable, but also in the interests of environmental quality. In the early days of the industry in each country, such chemicals

Movement of disease to a new area risks transfer to wild stocks with no natural immunity

have often been used freely, and with little consideration for the broader impacts. One of the dangers of the release of antibiotics into the environment is the emergence of resistant bacteria; another is the possibility of alteration of the bacterial flora in sediments affected by the discharges, because while most antibiotics break down rapidly in water, they persist longer in sediments. As the industry has matured however, national, European or US and international regulations have been developed and are increasingly being implemented.

5.5 Regulatory controls in aquaculture

All aspects of the aquaculture industry are regulated by national and international controls. For example, the main authorities in the US are the Food and Drug Administration (USFDA) and the Environmental Protection Agency (USEPA). In Europe, the Commission for mandatory implementation and application in all Member States issues the EU Decisions and Regulations, and EU Directives are issued by the EU Commission for incorporation into the legislation in all Member States.

In addition, the International Council for the Exploration of the Seas (ICES) has developed strict international protocols for handling and introducing exotic species; while the Office International des Epizooties (OIE) is a broader international organisation which promotes world animal health, and defines minimum health guarantees to be required of trading partners in order to avoid the risk of spreading aquatic animal diseases. These guarantees are based on inspection by competent authorities, epidemiological surveillance, and standard, intercalibrated methods for laboratory examinations and disease diagnosis. OIE has cooperative agreements with FAO, WHO,

Inter-American Institute for Cooperation on Agriculture, and Pan-American Health Organisation/World Health Organisation.

Thus, all steps in the aquaculture process are regulated, from the conditions of approval for the siting of a farm or hatchery, to the movement of fish and shellfish, to prevention of pathogens in feedstuffs of animal or fish origin, to monitoring for health and disease, to reporting of outbreaks of diseases of particular concern, to the quality, safety, efficacy and conditions of use of chemicals or drugs, licensed for use in aquaculture, to the monitoring of aquaculture products for chemical residues, to the disposal and processing of animal waste.

In conclusion, aquaculture is an industry, which applies biological, physiological and ecological principles. There is potential for adverse impact on the environment if aquaculture enterprises are poorly managed, as was seen in the early days of the industry in many countries. However in a mature, responsible industry, wherein national and international agreements, regulations and laws are adhered to and implemented, aquaculture would undoubtedly be sustainable (see Section 8.5).

Summary

- Parasites and disease are part of the natural biology and function of ecosystems.

- Diseases can be caused not only by micro- and macroparasites but also by non-infectious agents.

- The rearing of fish and shellfish under aquaculture conditions can cause stress, which can undermine their resistance, and make them more prone to disease.

- Diseases and parasites can be prevented or controlled by a variety of approaches including:

  - optimising growing condition;
  - breaking the life cycles of pathogens by appropriate management;
  - vaccination;
  - judicious use of approved medication.

- Pathogens, diseases and medications used to treat them can potentially impact on the environment:

  - movement of fish and shellfish stocks has resulted in some situations in the spread of diseases in the wild;

  - indiscriminate release of antibiotics in farm wastes can result in the emergence of resistant bacteria, or the alteration of the bacterial flora in the environment.

- The aquaculture industry is now regulated at both national and international level to ensure the sustainability of the industry.

Genetic effects on wild fish and invertebrates of accidentally or deliberately introduced cultured organisms

6.1 Introduction

Closed-cycle farming of both freshwater and marine fish and invertebrate species is increasing throughout the world (Section 1). Simultaneously, the number and extent of stocking/ranching exercises are also increasing, particularly with marine species. Deliberate (as in stock enhancement or ranching) or inadvertent release (as in farmed escapes) of cultured fish or invertebrate strains, within the natural range of a particular species, may have harmful effects on the genetics of wild populations of the same or closely related species. To minimise these effects it is necessary to understand their causes to allow optimal design of ameliorative measures.

Desirable qualities in farmed animals differ from those needed by wild animals

Direct genetic effects

Direct genetic effects are involved when interbreeding occurs amongst cultured animals in the wild, or between cultured strains and wild populations. The relative fitness of the resulting progeny is then in question. Most of the data currently available come from work on salmonid fish, particularly Atlantic salmon, of which more than 500,000 tonnes are currently produced annually by farms. However, the same basic principles are likely to apply to most aquaculture species.

The mechanism invoked for detrimental conspecific effects is as follows. Collection of brood stock, and subsequent artificial spawning and rearing of progeny, almost inevitably results in genetic changes to progeny, which lead to reduction in fitness in the wild. These phenomena are discussed in Section 6.2. When these cultured progeny are released or escape to the wild, some may breed with native conspecifics resulting in offspring of reduced fitness, thus reducing productivity in the area of interaction. The possible results of such interactions are discussed in Section 6.3, as will interspecific hybridisation between cultured strains of a certain species and wild members of related species.

Indirect genetic effects

Indirect genetic effects occur when there are ecological (see Section 7) or disease interactions (see Section 5) between released or escaped cultured strains and wild populations, resulting in drastic reduction in the size of wild populations. The reduced population may then become susceptible to a number of potentially detrimental genetic effects, as described in Section 6.2 below.

6.2 Brood stock and progeny manipulation

Methods of manipulating brood stock and progeny differ between fish or invertebrates intended for deliberate release to the wild (for stock enhancement or ranching), and those intended for closed-cycle farming. The objective in rearing for enhancement/ ranching should be to attempt to maintain similar genetic composition and level of genetic variability to those existing in the wild donor population. In contrast, it is never intended to

55

6

release farmed animals into the wild, so the imperative is to produce individuals that grow fast and have optimal traits for farming conditions and for marketing.

Principles common to rearing for deliberate release or for closed-cycle farming

There are, however, some common principles, which need to be invoked to minimise environmental impact. Within-population considerations will first be discussed, with problems relating to population structure being emphasised later. These principles relate to initial brood stock acquisition and the number of individuals used. Brood stock animals should be chosen from throughout the spawning season and not from one short period, as has often been the case in the past. The latter can lead to losing some components of the genetic variability present in a population.

In the past very low numbers of brood stock (less than 10 of each sex) have often been used to found a cultured strain. This was because of the very high fecundity of many cultured

aquatic species. However, as shown in Figure 15, this strategy will inevitably lead to large losses of genetic variability through inbreeding, which induces high levels of genetic drift. It will also lead to unpredictable changes in genetic composition, resulting in a loss of local adaptation (see below). (Artificial selection can also lead to changes in genetic composition, though in this case some changes are predictable).

Figure 15 also cites the 'effective number of parents' (N_e), which is the same as the actual census number (N_a) when the sex ratio is equal but moves rapidly towards the lower number sex, as the sex ratio varies greatly from equality. A figure of at least 50 individuals of each sex, equally-represented has been recommended by many authors, but others caution that this is a minimum number and that at least four times more brood stock are required even in the short term.

It is generally agreed that many thousand brood stock are required to maintain the evolutionary potential of a population. With

Brood stock for aquaculture should have as variable a genetic background as possible

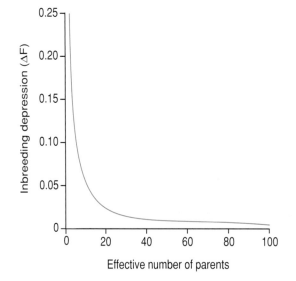

Figure 15. The level of inbreeding (expressed as ΔF, the coefficient of inbreeding), increases substantially as the effective number of parents decreases. A ΔF value of 0.25 means that 25% of existing genetic variability is lost in each generation, where the effective number of parents is two. The effective number of parents is the same as the actual number if the sex ratio is equal, but reduces drastically if one sex is greatly outnumbered by the other.

6

lower numbers of brood stock many rare allelic genes are lost from cultured stains, but the functional importance of such alleles has not yet been fully established. The current brood stock situation with two common farming species Atlantic salmon and Atlantic halibut are detailed in Info Box 3 and Info Box 4 respectively. While adequate control to prevent loss of genetic variability (beneficial if escapes occur) has been possible in salmon, it is clear that the current situation with halibut is unsatisfactory. Until the latter can be manipulated to optimise variability, escapes should be vigorously prevented and no ranching attempted. These examples illustrate that the level of risk associated with culture of a particular species can vary, and that the situation for each farmed species must be considered individually.

Info Box 3: Brood stock collection and progeny manipulation in Atlantic salmon farming: In the case of Atlantic salmon in the major producing country, Norway, adequate numbers of brood stock were taken from a large number of rivers and performance of the different riverine populations compared under common freshwater and sea farm conditions. The best performing six strains, in terms of growth, were then chosen, and have been used in a national breeding programme for more than 25 years. While this approach was initially adopted to optimise performance of the subsequent breeding programme, it also ensures that genetic variability is relatively high in farm escapees.

Info Box 4: Brood stock collection and progeny manipulation in Atlantic halibut farming: At present it is necessary to obtain wild brood stock in Atlantic halibut farming, since it is not yet possible on a commercial scale to rear both sexes of cultured halibut to sexual maturity and to achieve successful egg fertilisation. This is a large deep-water marine species reaching 3 m in length. Commonly, in the major farming countries of Norway, Iceland and eastern Canada, 20 to 30 large individuals, with approximately equal numbers of each sex, are taken by long-lining and transported to coastal locations, where they are maintained in large sea-water tanks or cages. Since halibut are portion spawners (i.e. they do not have all eggs ripe simultaneously), mesocosm spawning rather than stripping is used. Individuals are allowed to spawn naturally and their eggs, which are pelagic, are collected in the overflow to the mesocosm. Therefore, without using some genetic method of parental identification, it is not certain whether all individuals have spawned. The very high mortality that occurs during egg incubation potentially reduces N_e further. Thus cultured halibut are likely to be highly inbred compared with wild individuals, and escapes might be particularly damaging. In addition, future culture performance of farmed halibut may be compromised.

6

Rearing strains intended for deliberate release

Within population considerations: When dealing with strains intended for deliberate release into the wild there are several other points that should be considered. Failure to observe these strictures will mean that the damage to wild conspecifics may be increased. Artificial selection must be avoided (e.g. selection of certain brood stock such as the largest or earliest spawning; use of above ambient water temperatures during rearing to accelerate development) and efforts must be made to avoid relaxation of natural selection during culture. Equal numbers of each family must be available for release, otherwise the more abundant families will have a major influence amongst post-release survivors and N_e will be reduced.

Many authors recommend that each generation are the progeny of wild-caught brood stock, i.e. no cohort spends more than one generation in captivity. Such a strategy may be impractical if many generations of stocking/ranching are required, as is often the case when stocking or ranching is used to enable heavy fishing (so-called culture enhanced fisheries). However, keeping individuals for part of their life in a hatchery for several generations greatly increases the likelihood of genetic changes. Effective number of parents (N_e) must be similar for each generation of stocking/ranching, since even one generation with reduced numbers of brood stock inordinately affects average N_e.

A field experiment was undertaken in the West of Ireland in the 1990s as part of a European Union funded project (Figure 16). The aim

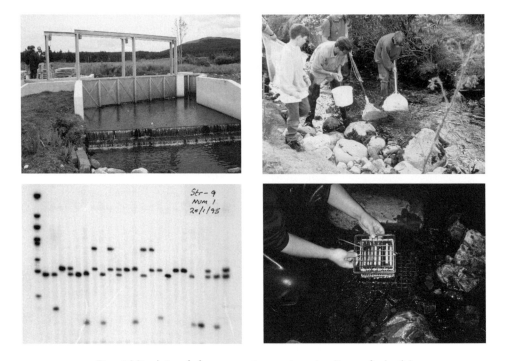

Figure 16. Simulation of a farm escape using genetic tagging. (See text for details.)

6

was to assess the relative performance in freshwater of farmed and native wild salmon, and their hybrids. A stream, with a natural spawning run of about 30 Atlantic salmon, had a trap installed, which catches all upstream and downstream migrants (top left). This trap is 1.7 km below an impassable waterfall. All native spawners were collected in early winter and were artificially stripped, as were equal numbers of farmed brood stock. Sixty families were produced; 15 wild, 15 farmed and 15 of each kind of hybrid (wild or farmed dam). All brood stock were screened for genotype at several highly variable minisatellite DNA loci (bottom left – where each slot represent the allele/s of a single fish, except the left hand slot which is a size marker). Eggs from all families were reared in the hatchery to eyed stage in the following March, then placed in the experimental stream in plastic mesh containers (bottom right). In early September 300 fry were collected by electrofishing (top right) and screened for the same DNA loci, allowing offspring to be identified to family. Significantly more wild than farmed offspring were present in the sample while hybrid numbers were close to expected values. When the entire experiment was repeated in the next year results were similar, thus proving the reduced fitness of farmed progeny in early life in freshwater.

As stated above, the objective when rearing for release, both to achieve long-term success and to minimise environmental damage, should be to avoid changing the genetic composition from that observed in the wild population from which the brood stock are derived. A comprehensive field comparison of cultured progeny and their wild progenitors, incorporating molecular genetics, is required, to ascertain whether this objective has been achieved (Figure 16). This type of survey has rarely been carried out in the past. It is

important to note that such a study will only directly detect changes in composition and level of variability of the genes being considered. These changes, however, are indicative of changes in the rest of the genome, as has been shown by comparative studies.

Between population considerations: It has been suggested in the past that it would be a good strategy to mix brood stock from several genetic populations to maximise variability, effectively producing a synthetic strain for stocking/ranching. More recent findings with anadromous salmonid fish contradict this suggestion, because it has been shown that a high level of local adaptation occurs in each population. Crossing of two or more locally adapted populations leads to a phenomenon termed outbreeding depression. This phenomenon usually does not affect fitness in the first generation of progeny (F1), because each homologous chromosome pair contains one copy from each population. Also, hybrid vigour can mask outbreeding depression by positively affecting fitness.

It is in the second and subsequent progeny generations (F2+) that severe reduction in fitness will be observed, as has been demonstrated with Pacific salmon in one recent study. Thus progeny of the crossing of two or more populations are likely to have reduced fitness in nature from the F2 generation onwards. When these individuals interbreed with wild fish they will in turn produce offspring of reduced fitness, leading to loss of productivity.

To avoid such effects it is suggested that single strains of local origin be used for enhancement/ranching. Where a population has become extinct and must be replaced, in what is termed a reintroduction exercise, it is recommended that the nearest viable

6

population is used as the donor, since there is often a positive relationship between genetic and geographical distance between populations.

In most marine fish and invertebrate species, populations are not as differentiated as in anadromous and freshwater species, so the extent of local adaptation is not, at present, clear. However, until experimental evidence on relative fitness of local and non-native populations is available for marine species it is recommended that the precautionary principle be invoked, i.e. single populations of local origin be used in stocking/ranching.

Rearing for closed-cycle farming

Within population considerations: In rearing for farming, it is never intended that progeny will be released to the wild (even if this actually

happens), so very different principles apply. The major farmed species are now the subject of breeding programmes where mass or family selection is applied to improve various commercially important traits. Agents of natural selection that apply in the wild are usually prevented from acting. For example, individuals are often cultured at constant elevated temperatures and non-ambient light regimes to shorten production time. Sex and ploidy (chromosome set number) may be manipulated (Figure 17). Since the females of many fish species grow faster and mature later than males, all-female strains are often produced. In oyster species and in rainbow trout, triploids (with three sets of chromosomes, two of which are of maternal origin) are often farmed since there is continued growth whereas growth slows in non-manipulated individuals when they mature sexually.

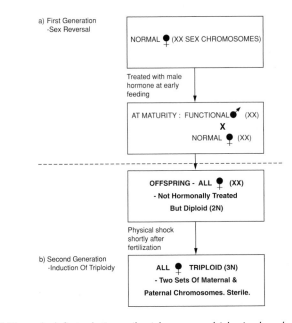

Figure 17. The methods for producing sterile rainbow trout and Atlantic salmon by sex reversal and induction of triploidy.

6

Finally, transgenic animals are beginning to appear for commercial utilisation. These animals are Genetically Modified Organisms (GMOs) in the sense that genes, usually for enhanced growth, are introduced from related species. All of these manipulations make farmed fish and invertebrates potentially less fit in the wild (see Figure 16). Since it is commercially vital to the industry to perform some of these manipulations, the strategy for environmental authorities should be to enact legislation to minimise or stop escapes, or to eliminate genetic interactions by farming only sterile animals. These strategies are discussed further below.

Between population considerations: Often, during the development of breeding programmes for farmed aquatic animals, different populations are crossed to maximise genetic variability. If such mixed strains escape from aquaculture facilities, they are likely to have lower fitness in the wild than single population strains (see above). Progeny of interbreeding with native animals are also likely to have reduced fitness. This provides another cogent reason for minimising or preventing farmed escapes.

6.3 Performance in wild and effect on native conspecific populations

Deliberate releases (stocking/ranching)
To minimise detrimental effects on wild conspecifics, while ensuring maximum success of a stocking/ranching exercise, it is important that only animals originating from the same population are used, i.e. native stocking/ranching. Where this principle is not observed, outbreeding depression effects (as described above) will lead to greatly reduced fitness of F2 and subsequent progeny of cultured x wild crosses. While the majority of supporting data

for this scenario come from work on anadromous salmonids in both the North Atlantic and Pacific (which exhibit genetic differences between populations that spawn in different river systems), it is likely that the same principles apply to fully marine species, even though the open nature of the sea enhances gene flow amongst populations. In the case of marine species, molecular studies demonstrate smaller genetic differences between populations than amongst salmonids, but genetic population structuring (i.e. differences) is still evident. It should therefore be assumed, until contrary experimental evidence is available, that any interbreeding between marine populations will be detrimental to the fitness of progeny.

It is important to reiterate that, from the perspective of minimising environmental impact, only fish or invertebrates from the local native population should be used in stocking or ranching exercises. Furthermore, modern molecular genetic investigations should be carried out on the natural population in an area and the cultured strain used for release, to assess the degree of detectable change in genetic composition and reduction in level of genetic variability. If statistically significant changes are evident in these parameters, detrimental environmental impact is highly likely, and the programme should be aborted or restarted with more attention to rearing and release protocols (see above).

Accidental releases (farmed escapes)
Even though the intention in closed-cycle farming is that there be no releases into the wild, escapes presently occur, particularly from salmonid sea cages/pens. Even an escape of say 1% of production (a figure often cited for Atlantic salmon farming, where farm production currently exceeds 500,000 tonnes)

6

means that 5,000 tonnes of escaped fish reach the wild each year (a tonnage that currently exceeds the entire North Atlantic commercial and sport catch). Many escapees are known to survive and join spawning runs into rivers, with incursions being heaviest in major farming areas such as western Norway, Scotland and Ireland.

Interbreeding with wild fish has been observed and the extent is evident from molecular genetic studies. Where field experiments have been designed to assess the relative fitness of farmed escapes and wild native salmon, farmed fish are found to have reduced fitness both in freshwater (Ireland – see Figure 16) and over the whole life cycle (Norway). In the Irish example, it was found that native fish were being displaced from the experimental stream by faster growing but lower surviving farmed fish. The farmed fish in question grew faster because they had been selected for fast growth over several generations. This phenomenon increases the rate of environmental degradation since farmed fish of reduced fitness rapidly replace native salmon.

Another problem in the Irish situation is that most of the farming industry uses one particular strain of Norwegian origin. The reason for this is that the largest longest-running breeding programmes for Atlantic salmon in Europe occur in Norway. Escapes of this single strain over a wide area break down the population structure of wild populations with which they interbreed. In consequence, fitness of progeny can be reduced due to loss of local adaptation and by outbreeding depression effects. It appears that little experimental work on the genetic effects of farmed escapes on wild conspecifics has been carried out with other fish and invertebrate species, but it is recommended that the findings with salmonids

be considered when determining policy, until experimental studies are undertaken on these other species.

There are two approaches to minimising genetic interactions between farmed and wild aquatic animals; either escapes can be completely prevented or farmed organisms might be made sterile so they cannot interbreed with wild conspecifics. Completely preventing escapes is difficult in species that are farmed in cages, because cages can be damaged by adverse weather conditions or other factors. The Norwegian authorities have been successful in reducing the proportion of escaped salmon by carefully matching the equipment to the exposure and wave climate of the particular site. However, the industry has been growing rapidly, so that in a time period where the proportion of escapes halved, production has approximately doubled resulting in similar numbers of salmon escaping.

It has been suggested, for example in salmon farming, that the industry be regulated to prohibit the use of floating cages and to use onshore tanks instead. However, production costs in tanks are substantially higher than those in cages, so for the industry to survive in the present economic climate, cages must continue to be used.

The other suggested strategy is to farm sterile animals. The only currently practical method of ensuring sterility in farmed salmonids is by inducing triploidy (Figure 17). In the case of rainbow trout, *Oncorhynchus mykiss*, sterile triploid individuals have been farmed for more than two decades, not to ameliorate the effect of escapes, but to prevent the reduced growth that occurs when normal diploid fish become sexually mature. However, with rainbow trout, fish must first be made all-female using hormonal treatment in the

Escapes are inevitable in cage culture. Even a 1% escape rate transfers huge quantities of genetic material to the environment

Escaped farmed salmon have damaged wild salmon populations in Ireland and Norway

6

generation prior to inducing triploidy, so as to ensure complete sterility (triploid males still undergo sexual maturation). Therefore, the process is protracted and expensive.

A similar process is feasible with Atlantic salmon (Figure 17), but the triploid progeny are more susceptible to husbandry stress, particularly in situations of reduced oxygen concentrations, so the industry has been reluctant to embrace triploid technology. Until simpler cheaper methods of inducing sterility, with no effect on performance, become available, it seems unlikely that they will be embraced by the aquaculture industry, unless forced by legislation.

Inter-specific hybrids

In the case of anadromous salmonids, cultured fish released to the wild appear to hybridise more freely with closely related species, than do wild fish of the same species. This may be because cultured fish have a lower position in

Escaped farmed salmon hybridise more easily with other species than do wild fish

the species breeding hierarchy, and thus have difficulty in securing mates of the same species, so spawn with a smaller species. With brown trout *Salmo trutta* and Atlantic salmon, hybridisation is considerably more common between wild brown males and farmed salmon females, than between wild fish of each species. Salmon x brown trout F1 hybrids are rarely fully fertile so there is little chance of introgression of the two species, but this may not be the case for all aquaculture animals, so this is another cause of environmental concern when cultured animals are deliberately or accidentally released to the wild. Even where F1 hybrids are of reduced fertility, hybridisation can be a problem. In western USA diploid rainbow trout have been stocked into waters occupied only by endemic cutthroat trout *O. clarkii*, with the aim of improving angling. Widespread hybridisation between the two species occurs, leading to the drastic reduction in population size of native cutthroat trout.

6

Summary

- The number of inadvertent (through farmed escapes) and deliberate (through enhancement/ranching exercises) releases of reared aquatic animals to the wild is increasing.

- Because captive breeding and rearing usually leads to reduced levels of genetic variability and differences in genetic composition compared with wild progenitors, such releases and subsequent interactions can reduce the fitness of wild populations, in many direct and indirect ways.

- The presence of many reared aquatic animals in the wild may also increase the incidence of hybridisation with closely related species.

- To minimise potentially detrimental effects Regulatory Authorities are advised to insist that enhancement/ranching exercises use only native or near native individuals as brood stock, enact rules to minimise genetic alteration during breeding and rearing, and prohibit mixing of populations as brood stock.

- In the case of closed-cycle farming genetic manipulation is regarded as essential by the industry, as in any other area of animal husbandry. Thus legislation should be enacted to attempt to eliminate escapes or to render farmed strains sterile, so that they do not interact genetically with wild populations, if they escape.

Interactions with wildlife

7.1 Ecological impacts of escapes and alien species

Genetic impacts of escapes have been dealt with in Section 6, while ecological effects of disease transfer were considered in Section 5. Here we are concerned with the consequences of deliberate or accidental release either of culture animals themselves, or of animals associated with them. In 1994 the FAO presented data that showed that 39% of all known introductions of alien/exotic aquatic species was related to aquaculture – and that 10% of aquaculture production was dependent upon introduced species. Many introduced species appear to cause little documented adverse ecological effect, or introduction took place so long ago (as in the case of the common carp, worldwide), that it is too late to determine effects. However, some escapes and releases have definitely caused ecological damage; a few examples are given here.

Deliberate release
Tilapias (*Oreochromis mossambicus*) have been introduced to many countries, as they are a mainstay of extensive and semi-intensive culture in tropical and subtropical areas. Generally they are regarded as a pest that reduces diversity – in the Philippines they have displaced the endemic *Misticthys luzonensis*, which has been brought close to extinction.

Over much of the world, the Japanese oyster *Crassostrea gigas* has become the mainstay of

Cultured species have been introduced to several ecosystems where they have out-competed native species

oyster culture. In most warm-water countries it has been accepted that it will breed and spread into the environment, competing with native bivalves, but in Northern Europe it has been normal to assume that the water temperatures are too low to permit breeding, so *C. gigas* has been reared from captive-bred spat (usually replacing the native oyster *Ostrea edulis*) with little or no thought of its spread. Whether because of acclimatization or global warming, it is clear that this assumption is faulty: breeding populations are now established in the Netherlands, displacing other bivalves, and there are signs of breeding activity in southern UK.

An under appreciated escape problem is that associated with the red-eared slider, a brightly-coloured freshwater turtle (*Trachemys scripta*) (Figure 18) from south-eastern USA. For many years the red-eared slider has been the mainstay of worldwide trade in pet chelonians, being bred on farms primarily in Louisiana. The trade was greatly boosted in Europe and the USA by the 'Mutant Ninja Turtles' children's craze of the late 1980s, but cultural/religious factors have also promoted strong sales in SE Asia, particularly Malaysia and Singapore.

Many sliders die in captivity, but the survivors grow to large size and are often abandoned by their owners. In cool temperate areas this has led to small non-breeding feral colonies (e.g. near Falmouth in the U.K.) that will die out eventually, but in warm temperate and tropical

7

areas, breeding populations have established themselves. Adaptable omnivores, sliders eat fish and invertebrates and have become top predators in many freshwater habitats, often out-competing local turtle species. They are often suspected of taking chicks of water birds, though direct evidence seems to be lacking. Ponds and lakes in public parks in Singapore are dominated by them, while freshwater bodies in Bermuda (a country plagued by introductions) have been heavily colonized over the last decade; there the sliders eat mosquito fish and a variety of local snails.

Accidental release
Shellfish, particularly molluscs, have been transferred around the world for well over a century as part of aquaculture initiatives. This has resulted in the transfer of several other animals that have 'hitchhiked' with them. A particular problem with bivalve shellfish is that they harbour commensal species (e.g. pea crabs, copepods) within their mantle cavities,

and have complex fouling communities living on their shells. They can also transfer phytoplankton bloom (that can cause toxic 'red tides') as resistant cysts in their gut contents.

A classic 'hitchhiker' example is the slipper limpet, *Crepidula fornicata*, a filter-feeding gastropod mollusc. This species was introduced to Europe from N. America in about 1880 with consignments of oysters. Extremely prolific, it became a major pest on natural and artificial oyster beds, smothering oysters either directly or by faecal deposition, and outcompeting them for food. It also invades mussel beds and has replaced native filter-feeders in several parts of the southern U.K. Another well-known pest species introduced in this fashion is the predatory American tingle *Urosalpinx cinerea* (a whelk), which directly preys upon young oysters. Not only animals have been transferred in this way: several alien seaweed species (e.g. *Laminaria japonica*) appear to have been

Figure 18. Red-eared slider (*Trachemys scripta elegans*).

Bird predation on cultured animals can create conflict between farmers and conservationists

introduced to Europe with oysters. Modern quarantine and inspection regulations should avoid this sort of transfer in the future.

7.2 Enhancing food supplies to wildlife

Food supplies, structures and artificial illumination of aquaculture facilities, attract wildlife. Farmers regard some of these wildlife 'guests' as pests because they feed directly on the species cultured or compete with them for food. Conversely, wild animals can become prey of the farmed stock. In extensive and semi-intensive aquaculture this additional food supply is generally desired but not in intensive systems, where the savings of feed costs are insignificant compared with the risk of infections with pathogens from wild populations. The interactions between wildlife and farmed animals are not restricted to predator-prey relationships and competition. The species attracted may mitigate the environmental impact of farming activities by consuming uneaten feed and faeces or increase the productivity of aquaculture operations by removing fouling organisms (e.g. seaweed, barnacles, mussels) and by helping to control parasites and predators.

High densities of fairly accessible prey make aquaculture facilities attractive to predatory water birds and aquatic mammals. These visitors can cause economic losses by stressing, injuring and consuming cultured species or by damaging nets – in the worst case liberating the whole stock of a cage unit. Fish stressed or injured by predators show reduced growth and are more susceptible to diseases. Furthermore, water birds complete the life cycle of many parasites of fish, shellfish and crustaceans, so birds can transfer diseases from farm to farm. Because many intruders are often warm-blooded animals that attract high public

sympathy, serious conflicts arise between conservation and ethical issues on one hand and the valid economic interests of aquaculturists on the other.

Eider ducks (*Somateria mollissima*) are attracted by mussel farms, where they find their favourite prey in abundance. One bird can remove about 2.6 kg per day. Bottom cultures of mussels are especially susceptible to invertebrate shellfish predators, i.e. shore crabs, starfish, and drills (marine gastropods that kills bivalves by drilling holes through the shell).

Ponds and raceways are susceptible to predatory birds from kingfishers to ospreys. Cormorants and herons can often cause significant losses of fish and crustaceans. Shallow ponds offer wading birds optimal foraging sites. There are also reports on predation by wild mammals, including river otters (*Lutra canadensis*), mink (*Mustela vison*), feral cats, bears and racoons. Predation by mammals, however, rarely poses a significant threat to freshwater aquaculture, perhaps with the exception of muskrats which can cause pond banks to collapse and dams to leak.

Salmon farms attract harbour seals (*Phoca vitulina*), grey seals (*Halichoerus grypus*), river otters, and sea lions (*Eumetopias jubatus* and *Zalophus californianus*).

There are several cases where aquaculture provides food for wild animals (e.g. Figure 9), to an extent that certainly influences the pattern of distribution of the animals and may in some cases increase their population size. Some of the better-documented examples are outlined below.

Eider ducks
Eiders are widely distributed sea ducks that inhabit coasts around the globe in polar and

subpolar regions, and specialise in feeding on mussels. They locate the mussels by diving in shallow water and pulling them off the substrate underwater, swallowing them whole and then grinding them up in their highly muscular gizzard. Farmed mussels grown on suspended rope cultures are particularly attractive to eiders because these tend to be thin shelled and have a high energy content compared with wild mussels growing intertidally. One bird can remove about 2.6 kg per day.

Rope cultures of mussels are particularly threatened because a high number of mussels are knocked off from the lines by the feeding ducks. Mussels are also grown at high densities and at shallow depths, so that energy costs of feeding are minimised for eiders allowed to feed undisturbed at mussel farms. Eiders, and to a smaller extent also some other sea ducks such as long-tailed ducks and scoters, soon develop the habit of flocking at unprotected mussel farms where they can rapidly deplete the standing stock of cultivated mussels. Experienced farmers tend to take steps to deter ducks from stealing cultivated mussels, but many farms do not protect their stock very effectively, and eider numbers feeding on farms can represent a high proportion of the local population. This is particularly the case in spring, when female eiders need to feed particularly intensively in order to build up reserves for egg production and for their fast throughout incubation. Somewhat unexpectedly, winter surveys of eiders feeding on natural habitat, at mussel farms and at salmon cages in the west of Scotland (Figure 19) found that eiders fed in very large numbers at many salmon farms as well as at mussel farms. In general, salmon farmers do not scare eiders away and the ducks evidently learn this as they often roost in close proximity to salmon

farms, where human disturbance may be reduced compared to other stretches of coastline due to salmon farms preventing human access in order to reduce risks of disease transmission.

It is not clear whether eiders feeding around salmon cages are simply stripping off mussels that foul the cages, nets and ropes, or whether eiders also scavenge lost fish pellet food. Certainly, captive eiders will happily feed on salmon feed pellets. Whichever is the case, there is no doubt that both mussel farms and salmon farms attract eiders, and have a strong influence on their local distribution from autumn to spring. Despite this strong, and relatively new, feeding association, there is no convincing evidence from detailed and accurate census data that the opportunities to feed at salmon and mussel farms has influenced population sizes of eiders in Scotland.

Shorebirds

Many species of shorebirds spend the winter-feeding on intertidal benthic invertebrates on estuaries and sheltered coasts. Where aquaculture is established on intertidal areas, high densities of cultivated animals may present unusually good feeding opportunities for some shorebirds. Oystercatchers in particular may benefit from being able to feed on mussels that have been artificially set on intertidal areas. Such opportunities are probably rather limited, as costs of disturbance will often outweigh the benefits from enhanced food stocks (see Section 7.3).

Gulls

Many species of gulls are highly opportunistic feeders, and exploit a wide range of habitats as well as taking many kinds of food. Where aquaculture permits gulls access, they will readily scavenge on remains of fish and

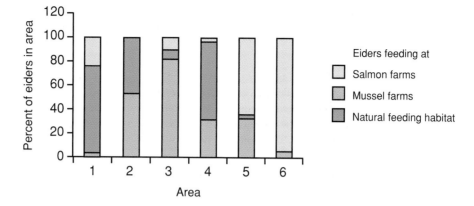

Figure 19. Distribution of eiders in six areas in the west of Scotland feeding at salmon farms, at mussel farms and in natural habitats in surveys in the autumn–winter of 1998 and 1999. Large differences between areas reflect differences in numbers of aquaculture establishments in different areas as well as variations in eider habits.

invertebrates, steal aquafeed pellets from feeders or by tearing open carelessly stored bags of feed, and take the growing product from accessible areas on farms. Gulls do not need to spend much time feeding each day, but can often obtain most of their daily energy needs in a few minutes of intense feeding at a place where food is abundant. It is very common to see flocks of several hundred gulls hanging around aquaculture facilities waiting for occasional opportunities to feed at the farmers' expense.

There is no doubt that the local distribution of gulls can be affected by the availability of such feeding opportunities, but populations of gulls use aquaculture as only a small part of their overall energy intake, and it is unlikely that this has much effect on gull population sizes. It is possible that the local aggregations of large scavenging gulls at aquaculture sites may drive away smaller birds that can be the target of robbery or predation by gulls.

Bird populations may benefit from aquaculture, just as from fisheries

Seals, cormorants, herons and other birds

Unprotected fish farms can attract seals, and a variety of fish-eating birds, especially cormorants and herons. The fact that these animals gather at fish farms implies that they find that these sites provide an attractive feeding opportunity. Often the animals attracted to fish farms are predominantly young ones, suggesting that adult animals perceive the hazards of feeding at fish farms to outweigh the benefits of access to easy pickings. Young animals are generally less efficient at foraging and are often displaced from the best feeding sites by older, dominant, animals. The inferior abilities of young animals may make fish farms relatively more attractive to them because they present a chance for rapid food intake, albeit at a high risk. Young animals at increased risk of starvation may find that feeding at fish farms increases their chances of overwinter survival providing they are not subject to high risks of being shot by the farmers.

7

7.3 Disturbance and wildlife persecution

Disturbance

Aquaculture can cause disturbance to wildlife by major modification of habitat, such as removal of mangroves required for nesting or feeding sites. It can also cause local disturbance through minor alterations to habitat. For example, oyster culture on intertidal areas involves addition of racks, stakes, culture bags, marker poles and other equipment onto open tidal flats. Some birds are attracted onto such structures. For example, gulls (and some kinds of shorebirds) may use elevated structures as roosts. Most species of shorebirds tend to avoid oyster culture plots, preferring to feed on open areas of tidal flats. Since shorebird numbers tend to be set by the amount of food in their wintering estuaries, loss of open estuarine foraging habitat to aquaculture is likely to have a negative impact on shorebird populations. This has not been studied in detail as yet. Aquaculture probably has much less impact on shorebirds than does loss of estuarine habitat through land reclamation or alteration of invertebrate populations due to nutrient pollution.

Human activity at aquaculture facilities can also affect wildlife. Where fish farms are sited in remote areas, as with many salmon farms or mussel farms in Scotland, wildlife may be affected by human disturbance resulting from routine farming activities in a way that animals would not be in places where they have become used to regular human activity. For example, black-throated divers nesting on remote lakes may be severely disturbed by helicopter flights transporting live fish into or from a farm, leaving their eggs exposed to predation by gulls or crows. Sea ducks may abandon otherwise profitable foraging areas if the level of disturbance by human activity increases their activity costs and reduces time available for foraging so that they are unable to balance their budget. Smaller birds, nervous of the potential threats presented to them by large gulls, may abandon an area in which large aggregations of gulls have come to gather as a result of feeding opportunities provided by aquaculture.

Acoustic disturbance

When properly used, acoustic deterrents have proved extremely efficient at deterring seals from predating from marine cages. Although some farmers have reported poor performance this has often been due to inadequate maintenance of equipment. There is not thought to be a significant danger or risk to seals as they appear to become acclimatised to the disturbance to some extent. There are, however, concerns that scarers may have a negative impact on cetaceans. It has not been possible to locate definitive research in this area, although it is clear that cetaceans are much more sensitive to these noises than seals. The effect of acoustic deterrents is related to their acoustic power and frequency range. A Canadian government study using a relatively powerful system showed that porpoises avoided approaching within a radius of 3.5 km from the scarer, while orcas were excluded from a much wider area.

Further research is needed on the effects on cetaceans of the variety of acoustic deterrents currently used, particularly since cetaceans are widely protected under international legislation.

Persecution

Many forms of aquaculture attract certain birds and marine mammals to feed on the high concentrations of food being cultivated, as discussed in Section 7.2 above. If no action is

Aquaculture in remote areas can cause disturbance to breeding birds

taken, birds and mammals can have devastating effects on the viability of farms. Farmers generally adopt one or more of three strategies. They may invest in costly structures to exclude wildlife from farms. They may reduce local numbers of the problem animals by shooting or other means. They may employ non-lethal deterrents that scare damaging animals away from the area.

Killing of wild predators of farmed fish is generally poor management practice. There is little scientific evidence that killing predators is effective in controlling predation. There is even less evidence that it produces an economic gain for aquaculture. By killing wildlife, the aquaculture industry paints an unflattering picture of itself as insensitive to the environment.

Some experts argue that there is little evidence that removal of bird or mammal predators has any long-term effect on predator abundance or fish loss at farms, because removed predators are quickly replaced by others attracted to such concentrations of food. Advocates of lethal control of predators argue that killing a few predators scares others and increases the effectiveness of non-lethal deterrent methods.

The main wildlife predation problems faced by aquaculture are fish-eating aquatic mammals and birds, especially seals, cormorants and herons. Marine salmon farmers in the United States estimate that 10% of their US$50 million annual farm gate value (i.e. US$5 million) is lost to seal predation. Some American salmon farmers argue that it is unfair that they cannot kill seals when farmers in the western United States can have government agents to kill wolves that harm their livestock, even though wolves are listed under the Federal Endangered Species Act. In contrast, seal numbers on the coasts of the United States

Seals eat up to 10% of marine cultured salmon in the USA

are increasing, to a large extent due to their protected status.

A survey of fish farmers in north-central states of USA found a strong consensus that farmers should be allowed to kill birds on their property without permits, and an unwillingness to invest money in preventative measures. The United States Fish and Wildlife Service reported killing of about 10,000 birds per year under permit in the early 1990s, mostly double-crested cormorants (*Phalacrocorax auritus*), herons and egrets. These birds were killed to reduce impacts of fish-eating birds in fish farms. Mississippi Delta catfish farmers estimated loss to cormorants of fish worth US$3.3 million per year, despite shooting birds and spending US$2.1 million per year on deterring birds. However, although the last example indicates large financial costs, these represent only a few percent of the industry's production value.

Predator control occurs in association with aquaculture in most parts of the world. In Southeast Asia, wildlife is protected in many countries, but licensed and illegal killing of birds at aquaculture facilities certainly occurs very widely. There are even reports of some fish farms setting monofilament nylon 'mist nets' over fish cages and ponds in order to trap birds attracted to the site. These birds may remain hanging in the nets until they die of exposure or starvation. The logic of this seems to be that any bird becoming tangled in these nets must have been attracted to the site by the prospect of stealing food, and therefore that death of these birds will benefit farm productivity.

Two studies of predator control at Scottish salmon and trout farms in the late 1980s estimated that around 500 herons, 1600 cormorants and 1400 shags were being killed

7

Significant levels of persecution of aquatic birds are associated with aquaculture

each year in that rapidly developing industry. The majority of these birds were killed illegally, only a small proportion being killed under license from the Scottish Office.

What effect does predator control have on bird populations, and can these impacts be justified by the economic gains that result? Data to answer these questions do not exist for most parts of the world. Even in countries where wildlife conservation is given a high priority and populations are carefully monitored, the facts are far from clear. Surprisingly, despite issuing licenses to permit farmers to kill birds, the USFWS has not made any definitive evaluation of the effect of shooting on population trends of cormorants, herons and egrets in the USA. There are suggestions of declines in cormorant numbers in Maine and of great blue heron numbers in Midwest and West USA areas associated with aquaculture and persecution of birds that are causing problems at farms.

Numbers of great cormorants in northwest Scotland have certainly declined in recent decades whereas in every other part of Europe their numbers have been increasing. It is likely, though far from proven, that the decrease in their numbers in northwest Scotland is a direct result of the killing of large numbers of cormorants at salmon farms.

Mussel farmers in Scotland shoot some eider ducks, and some are drowned as a result of becoming entangled in anti-predator nets set around mussel farms. However, numbers of eider ducks in Scotland are increasing in all areas except Shetland (where deaths of ducks associated with mussel farming have been negligible up to now), and there is no indication from census data that the rate of

increase has been noticeably reduced by the mortality associated with mussel farming in other parts of Scotland.

If it is difficult to establish the impact of, or even the extent of predator control by aquaculturists in the United States or Scotland, it is far more difficult to assess the situation in parts of the world where the lobby for wildlife protection is less strong, and where there is little monitoring of bird or aquatic mammal populations.

In addition to the impact of lethal control measures, non-lethal measures may affect wildlife. In this case the effects are probably minor and mainly involve impacts on local distribution and behaviour rather than on population size. Non-lethal scaring measures regularly used in the aquaculture industry include shooting with blanks, use of gas cannons, other acoustic deterrents, scarecrows, chasing by powerboat, flashing lights and pyrotechnics. Acoustic devices used to reduce seal activity at fish farms may drive away other marine mammals too. Porpoise and whale numbers are reported to be reduced within 3.5 km of these devices.

Regular scaring of birds from aquaculture sites may increase their energy needs through increasing the time they have to spend flying, and may reduce their longer term foraging rates. Sea ducks in areas with extensive mussel farming show higher rates of vigilance than ducks in areas without mussel farms, and respond more rapidly (by flying away) when approached by boat. This relates to the frequent habit of farmers to chase sea ducks away from mussel farms by powerboat. Birds learn to be more cautious in places where they are subject to such harassment.

7.4 Reduction of pressure on wild stocks

Food fish farming

From an economic point of view it might be expected that farming of over-fished species will relieve wild populations. With increasing supplies of cultured fish, prices should drop and the corresponding capture fishery on overexploited stocks should be reduced as margins shrink. An economic model from the 1980s, describing market interactions between aquaculture and common-property commercial fishery predicted that the entry of a competitive aquaculture sector would increase supply, reduce prices, and increase natural fish stocks. With the recovery of natural stocks the supply from capture fisheries should also increase. However, if the aquaculture sector became dominated by a small number of companies the model predicted a less benign impact on the situation of the wild stocks.

The real world fishery economy is more complex. For wild fish the market price can be several fold higher than for their farmed counterparts, and if fishermen are state-subsidised, as is common in many parts of the world, the revenue from the catch does not even need to compensate for the fishing effort. In addition, many fisheries do not target single species, so a reduced market price for a rare species does not necessarily cause a reduction in fishery effort. Furthermore, to describe the impact of aquaculture on wild fish stocks properly, a model would also have to take into account the negative ecological impacts of aquaculture identified in other sections of this volume. Of particular concern is the transfer of diseases to, and detrimental genetic impacts upon, the wild relatives of cultured food fish in the long run.

Predictions that farming fish reduces pressure on wild stocks are over-simplistic

In considering these complicated interactions between aquaculture, fisheries, and wild fish populations it is not easy to find an answer to the question of whether aquaculture fish production reduces pressure on wild fish stocks or not. An adequate modelling approach would require the integration of ecological and economic expertise in a manner that is presently in its infancy.

While it is difficult to verify a positive effect on wild stocks from ongrowing of food fish, it is evident that wild fish stocks may benefit from well-designed stock enhancement programmes, taking into account genetic, ecological, and behavioural peculiarities of the fish stock to be supported. Most stocks of Atlantic salmon (*Salmo salar*) and sea trout (*Salmo trutta trutta*) depend on releases of fry produced in hatcheries. The same is true for various species of sturgeon (*Acipenser* spp.) in Asia, Europe, and North America. Artificial breeding of endangered fish species may be the only chance for their conservation or restoration. However, in principle, stock enhancement programmes should be regarded as second choice substitutes for appropriate fishery management actions such as habitat restoration and reduction in fishing effort.

Culture of aquarium fish

The worldwide aquarium trade, predominantly for fish, is large and increasingly well organised and commercially valuable. Freshwater fish, both tropical and temperate have been held in aquaria for centuries and the consumer taste for fancy varieties (e.g. of goldfish, guppies, koi carp, cichlids and Siamese fighting fish) has led inevitably to captive culture that has taken pressure off wild stocks. The situation is different for marine fish, particularly from the tropics. Keeping tropical marine fish in aquaria

7

Captive breeding of aquarium fish is desirable ecologically

is a relatively new mass pastime (around 30 years old) as a number of technological obstacles had to be overcome to make this feasible in the home. Also, the great variety of colourful fish available in the sea means that there has so far been little demand for varieties enhanced by selective breeding. In consequence, the trade still relies heavily on capture of wild fish. Many of the techniques used (e.g. cyanide poisoning of coral reef fish) are extremely damaging and wasteful – as well as killing non-target fish, those fish that are captured often succumb to stress soon after capture, maintaining demand and perpetuating such practices. However, initiatives to rear marine fish locally in developing countries, together with mass culture of popular species such as clown fish, are changing the situation. Captive-bred fish normally survive far better in captivity that do wild fish that harbour disease and are stressed by confinement. Consumer demand for healthy, long-lived aquarium fish is likely to favour captive breeding.

Lobster stock enhancement

During the 1980s consideration was given to culture of lobsters, particularly the European lobster *Homarus gammarus* (see also Section 2). Lobster rearing to marketable size proved to be uneconomic. Instead, attention turned to stock enhancement, with small juvenile lobsters tagged with magnetic wire tags being released directly into suitable habitat (rocky reefs and cobbled sea bed). Survival rates have been high and tagged lobsters have recruited to lobster fisheries. This approach, coupled with local initiatives such as carapace V-notching (marking mature lobsters to disqualify them from the fishery) to reduce fishing pressure, has great potential for augmenting natural stocks of lobster – though currently poaching and resource ownership issues are impeding progress.

Summary

- Nearly 40% of all known introductions of alien or exotic species to aquatic ecosystems have been related to aquaculture.

- Many introductions have little known ecological effects, but others have caused extensive ecological damage.

- Aquaculture often attracts wildlife that benefits from extra available food, but can pose problems for the industry, which sometimes responds by destructive persecution. Examples of such wildlife include eider ducks, otters and seals.

- Physical changes in habitat resulting from aquaculture (e.g. destruction of mangroves, building of support structures and cages) can impact negatively on wildlife.

- Some aquaculture techniques (e.g. culture of aquarium fish, lobster stock enhancement) can take pressure off wild stocks.

8 Sustainability of aquaculture

8.1 Is current aquaculture practice sustainable?

'Sustainable development is the management and conservation of the natural resource base and the orientation of technological and institutional change in such a manner as to ensure the attainment and continued satisfaction of human needs for present and future generations. Such sustainable development (in the agriculture, forestry, and fisheries sectors) conserves land, water, plant, and animal resources, is environmentally non-degrading, technically appropriate, economically viable, and socially acceptable.'

(Code of Conduct for Responsible Fisheries, Food and Agriculture Organization of the United Nations, 1995)

The above definition demonstrates that 'sustainability' is of limited validity as a useful term, as it is susceptible to different shades of interpretation, particularly by politicians and economists who are primarily concerned with economic sustainability. Here we assess whether aquaculture is compatible with the sustaining of natural environments, ecological systems and biodiversity, at a range of scales.

It is undoubtedly the case that many past and existing aquaculture practices are incompatible with ecological sustainability. As a general rule of thumb, almost all large-scale land-based aquaculture (e.g. carp culture) involves

virtually total replacement of diverse natural terrestrial and freshwater ecosystems with artificial mono- or poly-cultures, often (as in China and other Asian countries) combined with intensive agriculture, particularly for rice. Over human timescales (hundreds of years) such environmental modification is essentially irreversible, and often unavoidable given human population pressure.

Much of coastal aquaculture is also ecologically unsustainable as presently practiced. Tropical and semitropical coastal pond culture, especially in mangroves, has been a particular disaster, and some countries and aquaculturists themselves have begun to back away from creating more of this type of damage. However, as with unrestrained deforestation, there are still countries (such as Indonesia) where short-term financial benefits mean that irreversible mangrove clearance for ponds will continue for the foreseeable future (Section 3).

Sea cage culture for fish such as salmonids in temperate waters and seabass in warmer waters is now global in extent. Generally it has fewer demonstrable direct sustainability problems, though the case that salmon culture causes increased sea lice infestations in natural salmonid populations is persuasive if not certain, while serious reservations remain about the genetic consequences of escapes of farmed fish (whether genetically modified or not) (Section 6). Certainly the available evidence suggests that such farming probably does not

Land-based aquaculture has similar effects to intensive agriculture

Mangrove clearance for pond culture is irreversible

8

Fishing and aquaculture interact
to remove largest and smallest
fish from ecosystems

have prolonged localised consequences as neighbouring pelagic and benthic communities revert to normal within a few years after farming is abandoned (Section 4).

However, the farming of such carnivorous fish does have indirect ecological effects. The distinguished fisheries' biologist Daniel Pauly developed the concept of 'fishing down aquatic food webs' in the late 1990s. By this he meant that, as top predators have been removed by fishers, there has been a consistent and measurable tendency to increase capture rates of fish lower in food chains. More recently he has sounded warning bells that aquaculture in much of the developed world (but not in the USA, Asia, Africa and the former Soviet Union) is now demonstrably 'farming up food webs' to sustain the demand of the global aquaculture feed industry (Figure 20), and is still heavily reliant on fish meal (see Section 4). Instead of taking only small planktivorous fish such as capelin, industrial fishers are now taking larger fish, higher in food chains as well. Sophisticated multi-species' fisheries modelling, using factors such as mean trophic

level of fish landed (see Info Box 5) can detect such two-way 'squeezing' of fish resources (Figure 21), but effective control of fishing pressure is difficult to attain despite the efforts of governments and international organizations world-wide.

Squeezing of fish resources is a serious problem for the aquaculture industry as well as aquatic ecosystems. Globally, feed manufacturers are attempting to find more ecologically sustainable protein and lipid sources than industrial fish (see Section 2), because of rising fishmeal prices. More precisely designed feeds, and an increased reliance on vegetable input (e.g. soya), are evidence of their technological success. However, just as capture fisheries have repeatedly been threatened and damaged by advances in gear technology, so problems arise from more technically efficient aquaculture.

Globalisation of retailing, coupled with effective marketing and an increasing emphasis on reliable sourcing of high quality food products is relentlessly driving up demand for formulated feeds that can deliver uniform

---

Info Box 5: Estimation of trophic levels (TL): Trophic levels are based on the position of organisms within food webs. A TL of 1 indicates plant or detrital material, a TL of 2 an herbivore or detritivore, a TL of 3 a consumer of herbivores and so on. The TL of fish usually falls between 2 and 5, though values of 5 are rare in fish, and more applicable to top carnivores such as orcas. In addition, most fish actually have a variety of

food items in their diets, so their trophic levels have non-integer values.

TL for a given consumer fish species (i) is given by:

$$TL_i = \sum_j TL_j \cdot DC_{ij}$$

Where TL_j = the fractional trophic levels of the preys j and DC_{ij} = the fraction of j in the diet of i.

8

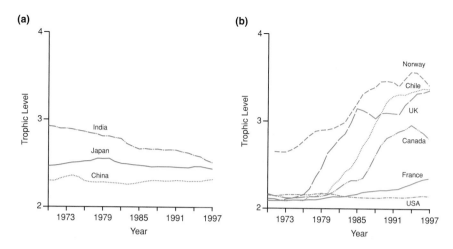

Figure 20. Contrasting trophic level trends amongst major aquaculture countries.
(a) Asian countries. (b) Western countries. (Redrawn from Pauly *et al.* 2001.)

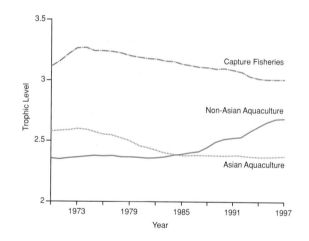

Figure 21. Two-way 'squeeze' of capture fisheries and non-Asian aquaculture on fishery resources.
(Redrawn from Pauly *et al.* 2001.)

desirable, tradable products. For example: a decade ago, milkfish in the Philippines would have received no artificial diets and would have been reared in extensive conditions where they would have been browsing at a low trophic level and marketed relatively locally. Now sea-cage manufacturers are promoting coastal sea cage culture of Philippine milkfish fed on pelleted diets to supply international markets.

Even in the absence of such high technology infrastructure, herbivorous and omnivorous aquaculture species (such as milkfish and tilapias), formerly fed on plant material, are increasingly being fed upon pelleted diets to improve their growth rate. Such is the growth of aquaculture worldwide, that any fishmeal saving generated by better diet formulations, will inevitably be offset in the short and

8

medium term at least by the faster-growing absolute demand for industrial fish.

There have to be serious reservations about the long-term ecological sustainability of aquaculture practices that are so crucially dependent on industrial capture fishing, particularly if those countries that currently practice low-input aquaculture switch increasingly to aquafeed-based operations.

There are possible comparisons with the BSE crisis in cattle, which was caused by an increased use of foodstuffs from higher up the trophic pyramid.

8.2 Ecological footprints

A recent approach to considering the question of sustainability in aquaculture stems from the general concept of the 'ecological footprint' – a measure of the sum of 'goods and services' that any human activity takes from the natural environment. In aquaculture this can be thought of as the area of ecosystem required to

supply food and oxygen to the culture facility, plus the area required to disperse waste products (though these areas may overlap or be very far apart). Table 4 gives examples of foot print areas calculated for different types of marine and freshwater culture.

The concept is at an early stage of development, and it is clear from Table 4 that the area of ocean required to deliver the fishmeal needed for diets (as in Baltic salmon farms and intensive tilapia culture) swamps all other considerations. However, it is evident that high technology, high input aquaculture has a considerably larger ecological footprint than extensive artisanal aquaculture.

8.3 Are herbivorous species the answer?

Fish culture
Most fish culture in the world is for herbivorous or detritivorous species. Farming of carp, milkfish and tilapias in Asia and channel catfish in the USA is generally

Table 4. Ecological footprints of different types of aquaculture. (From Black, 2001.)

Culture type	Area of ecosystem support required per unit farm area
Baltic salmon farms	40,000-50,000
Intensive tilapia cage culture in Lake Kariba, Zimbabwe	10,900
Prawn pond culture, Columbia	35-190
Baltic mussel longline culture	20
Semi-intensive pond culture in Lake Kariba, Zimbabwe	1

8

environmentally neutral (i.e. the 'ecological footprint' is small; see Section 8.2), save that it takes place in terrestrial and aquatic systems that are physically modified in consequence. Energy input is solar or results mainly from plant breakdown or addition of pellets with little inclusion of fishmeal and oil (4% in the case of channel catfish) and there is often (but not invariably) relatively limited input of nutrients to cause eutrophication. Much of such culture has proceeded successfully for centuries, though enhanced demand because of exponentially rising human populations and a trend away from local subsistence towards international trade has resulted in far more land being turned over to such use.

Bivalve culture

Culture of 'herbivorous' oysters, mussels, clams and scallops is usually regarded as a particularly benign form of aquaculture, especially as energetic inputs are usually limited because they feed on food entrained in the seawater around them (though spat may be reared intensively using cultured algae in some high value species). Bivalve molluscs, if not cultured on the bottom in fairly natural conditions, are suspended in or from a variety of line, cage, trestle and rack constructions in the water column above the seabed (Section 3).

In intensive systems this effectively creates a non-natural 3-dimensional suspended bivalve bed. Bivalves filter feed (predominantly on phytoplankton) by pumping large quantities of water through their mantle cavities (large natural mussel beds pump millions of tons of water per day) and extracting most particulate material and some dissolved organic materials, particularly amino acids. It should be noted that their gills are highly specialised for pumping water for feeding rather than respiration – they extract relatively little of the oxygen (*ca* 5%). In deep fjordic systems (e.g. in Chile, New Zealand, and northern Europe) few problems arise, though hydrodynamic modelling may often be needed to determine whether the intensity of farming is within the carrying capacity of the environment.

In shallow bays or lagoons (e.g. in France and Spain), farmed mussels or oysters can reach densities where they may have severe impacts on environmental carrying capacity. Within the water column they act as competitors for zooplankton in exploiting phytoplankton.

Intensive bivalve culture also has effects on the seabed as well as on the water column. Assimilation efficiencies in filter feeders are generally not high; so much of the energy intake of seston and larger particles will eventually result in the expulsion of faecal material with a significant organic content. Some of this is probably recycled within the seston, but much falls to the seabed. Effectively the bivalve culture extracts water column organic material and transfers it to the seabed. In shallow lagoons and bays this often results in enrichment of sediments, sometimes to the extent that macroalgal blooms (e.g. of sea lettuce, *Ulva*) cover the seabed.

Naturally sandy substrata become choked with fine material and a high BOD can cause anoxic conditions (see Figure 11). Such 'souring' of the seabed has in the past resulted in a high rate of abandonment and relocation of bivalve farms, though increasingly sophisticated modelling of carbon flux should make this less common. Overall, with greater care over site selection, minimization of transfer of hitchhiker species and disease, plus more sophisticated modelling of impacts, there seems little reason to doubt that ecologically sustainable bivalve culture is possible.

8.4 Does advanced technology provide solutions?

Advanced technology in aquaculture currently falls into three broad categories. First, the microelectronics and information technology revolution of the last forty years, combined with improved scientific knowledge of the physics, chemistry and biology of aquatic systems, means that the scope for understanding and predicting the ecological consequences of aquaculture operations has been greatly enhanced. This is obviously a positive technological benefit as far as ecological sustainability is concerned.

Second, aquaculture engineering has advanced considerably, so that sea cages can be massively larger and moored in far more exposed waters than hitherto. Conversely, closed-cycle, land-based facilities of great complexity have become feasible, permitting close control over farmed stock, feed supply and waste disposal, based on recirculation of water, whether fresh or marine. These developments have both positive and negative aspects. They clearly underpin much of the 'luxury' end of the aquaculture spectrum, so tend to increase demands for inputs of energy and aquafeed. On the other hand, they tend to reduce pollution and disease problems significantly, at least in the case of closed-cycle operations – the openness of cage culture still poses problems.

Finally, integrated land-based polyculture can provide a model for minimising the environmental impact of mariculture, even without recirculation. For example: in Israel a commercial farm based upon recent research combines culture of seaweed (*Ulva, Gracilaria*), the herbivorous gastropod abalone (*Haliotis* sp.) which feeds upon the seaweed, and pellet-fed sea bream and penaeid prawns, whose particulate effluent supports mullet

Modelling, plus enhanced scientific knowledge promote prediction of ecological sustainability

(*Mugil*) and sea cucumbers, while the dissolved nutrients are delivered to the seaweed culture units (Figures 22 and 23). Such systems can deliver an innocuous effluent to the sea, but require considerable sophistication of control and monitoring systems as well as much biological information and careful modelling to permit design of the tank sizes and biomass of each species contained therein – they also require appreciable areas of land, with consequent impacts upon terrestrial ecosystems.

8.5 What is the role of regulation in delivering sustainability?

Many of the aspects of the aquaculture industry in developed and developing countries are regulated by national and international controls (see Section 5.6).

Unfortunately, even in developed countries monitoring and enforcement of regulations is not perfect and some recent disease outbreaks, together with use of unlicensed chemicals, have been traced to evasion of regulations – just as sometimes happens in the related industry of agriculture. In some developing countries regulation is often much less rigorous and site planning in particular is often far less controlled than in the developed countries. In heavy industry and agriculture there has already been a tendency to export pollution and environmental degradation from the regulated developed world to the poorer developing countries; there are signs that this is happening too with aquaculture.

A particularly intractable regulatory problem worldwide lies in controlling industrial fisheries for the ingredients of aquafeeds, and coastal fisheries for brood stock. Essentially this is a subset of the general difficulty of controlling capture fisheries. Bluntly, few of

8

Figure 22. Integrated marine farm at Mikmoret, Israel, culturing fish, abalone and seaweed
(see Figure 23 for diagrammatic layout). (M. Schpigel.)

the world's fisheries are managed and regulated in an ecologically sustainable fashion at present; those fisheries that underpin aquaculture as presently practiced are no exceptions.

In a mature, responsible industry, wherein national and international agreements, regulations and laws, based upon the best available information, are adhered to and implemented, aquaculture would be

National and international
legislation, backed by effective
enforcement are essential

Figure 23. Diagram of integrated polyculture
farm in Israel (see text for details).

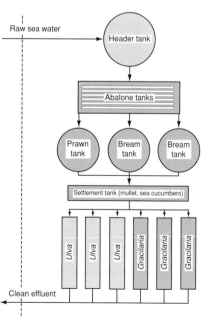

sustainable. Even though great progress continues to be made, we have some way to go before this is globally true.

8.6 Integrated resource management – an urgent necessity?

Aquaculture is not an isolated activity. As we have shown in this volume, aquaculture interacts with other human activities, in particular with fisheries and agriculture, the older and greater sources of food for humans. It is easy to pillory some of its practices, but its deficiencies cannot be dealt with in isolation either. At regional and local levels an integrated approach is needed. Too often

aquaculture, agriculture, fisheries, forestry, building development and tourism pull in different directions to the inevitable detriment of the environment and the ecosystems within it.

On a grander scale, the great problems of sustainability associated with aquafeed production are shared with intensive agricultural production of pigs and poultry, while the industrial fishing that delivers fishmeal to both industries is not managed more effectively at present than the rest of capture fisheries. There is clearly a need for integrated resource management at a global level.

Summary

- Capture fisheries and increasing use of industrial fishing to support aquaculture are squeezing fish resources in a fishing down, farming up fashion.

- Application of the concept of the 'ecological footprint' of various types of aquaculture is a promising but undeveloped approach to evaluating impacts.

- Culture of herbivorous fish and invertebrates, while generally more ecologically benign than culture of carnivorous forms, can nevertheless have substantial effects on ecosystems, particularly in large-scale intensive operations.

- Advanced technology brings many benefits – but also tends to raise demand and input of energy and aquafeeds.

- Regulation has a strong role in delivering sustainability – but enforcement remains a problem.

- Integrated resource management is essential to minimising environmental degradation.

9 Further reading

Black, K.D. & Pickering, A.D. (1998). *The Biology of Farmed Fish*. Sheffield Academic Press, Sheffield.

Black, K.D. (ed.) (2001). *Environmental Impacts of Aquaculture*. Sheffield Academic Press, Sheffield.

Bower, S.M., McGladdery, S.E. & Price, I.M. (1994). Synopsis of infectious diseases and parasites of commercially exploited shellfish. *Annual Review of Fish Diseases*, **4**, 1-199.

Bruno, D.W. *et al.* (1997). *What Should I Do? A Practical Guide for the Marine Fish Farmer*. EAFP (European Association of Fish Pathologists).

Burnell, G.M. (1996). The environmental impact of marine bivalve mollusc exploitation: a brief review of the disturbance caused by mariculture and fishing. In: Giller, P.S. & Myers, A.A. (eds). *Disturbance and recovery of ecological systems*, 84 - 97. Royal Irish Academy, Dublin.

Costello, M.J. & Boxhall, G.A. (eds) (2000). Proceedings of the Fourth International Conference on Sea Lice, 28-30 June 1999. Special Issue: *Aquaculture Research* **31**, 793-883.

Eleftheriou, M. (ed) (1997). *AQUALEX. A Glossary of Aquaculture Terms*. John Wiley & Sons, Chichester.

Elston, R.A. (1990). *Mollusc Diseases: Guide for the Shellfish Farmer*. Washington Sea Grant, University of Washington Press.

Naylor, R.L., Goldburg, R.J., Mooney, H., Beveridge, M., Clay, J., Folke, C., Kautsky, N., Lubchenco, J., Primavera, J. & Williams, M. (1998). Ecology – Nature's subsidies to shrimp and salmon farming. *Science,* **282**, 883-884.

Naylor, R.L., Goldburg, R.J., Primavera, J.H., Kautsky, N., Beveridge, M.C.M., Clay, J., Folke, C., Lubchenco, J., Mooney, H. & Troell, M. (2000). Effect of aquaculture on world fish supplies. *Nature*, **405**, 1017-1024.

OIE (Office International des Epizooties). (2000) *International Aquatic Animal Health Code.*

Pauly, D.P., Tyedmers, R., Froese & Liu, L.Y. (2001). Fishing down and farming up the food web. *Conservation Biology in Practice,* **2(4)**, 25.

Tett, P. & Edwards, E. (2002). *Review of Harmful Algal Blooms in Scottish coastal waters*. SEPA, Stirling.

Schlotfeldt, H.J. *et al.* (1995). *What Should I Do? A Practical Guide for the Freshwater Fish Farmer*. EAFP (European Association of Fish Pathologists).

9

Wackernagel, M. & Rees, W. (1996). *Our Ecological Footprint: Reducing Human Impact on Earth.* New Society Publishers, Gabriola Island, Canada.

Watson, R. & Pauly, D. (2001). Systematic distortions in world fisheries catch trends. *Nature,* **414**, 534-536.

Wildish, D.J. & Heral, M. (eds) (2001). Environmental Effects of Aquaculture. 2001. *ICES Marine Science Symposia Vol 213; ICES Journal of Marine Science 2001,* **58(2)**, 363-530.

10 Useful web sites

http://www.american.edu/TED/CRAYFISH.
HTM#IDENTIFICATION

http://www.ecoserve.ie/projects/sealice/

http://www.iffo.org.uk/

http://www.fishbase.org/

http://www.ramsar.org/

http://www.fao.org/

11 Glossary

Aerobic respiration: the oxidation of organic matter producing energy using oxygen as the oxidant.

Algal blooms: a build up of algal cell numbers that considerably exceeds natural levels.

Anadromous: an aquatic species which breeds in freshwater and moves to the ocean to feed (e.g. salmon).

Anaerobic respiration: the oxidation of organic matter to produce energy using a variety of oxidants such as iron, manganese or sulphate rather than oxygen. The reduced forms of these oxidants will subsequently be reoxidised if sediments become more oxidising e.g. if the organic supply is reduced.

Antibiotic: chemicals that can arrest the proliferation of bacteria, thus reducing disease. These are widely used in aquaculture.

Antifoulants: chemicals used on submerged structures to reduce the build up of fouling organisms. Toxic tin-based products have been superseded by less toxic copper-based products.

Antiparasitics: fish suffer from a variety of parasites. Sea lice are treated by a variety of medicines either administered in dissolved form in a bath or in mixed in with the feed. Biological control with 'cleaner' fish is also used.

Artificial selection: selection, applied by man, of specific individuals from a larger possible number, as parents for the next generation.

Berried: female crustaceans in an egg-carrying stage with eggs visible externally as projections from the underside of the body.

Brine shrimp: a crustacean (*Artemia* sp.) characteristic of salt lakes, about 1 cm in length as an adult, whose cysts (resistant, highly protected eggs that can be held dry for years) hatch into smaller larval stages that are fed to young fish and prawns as a live feed.

Brood stock: animals that are used as breeding parents to obtain young stages for aquaculture.

Conspecific: individuals of the same species.

Ecological footprint: the area of ecosystem required to supply food and oxygen to an aquaculture facility, plus the area required to disperse waste products.

Effluents: unusable matter coming out from a place such as a fish or prawn farm.

Endemic: found naturally in a locality.

Enhancement: stocking to increase numbers of a species, by augmentation of wild individuals.

Eutrophication: the unwanted stimulation of primary production caused by the release of nutrients from human activities.

Evolutionary potential: having sufficient genetic variability to undergo future genetic change.

FAD: an acronym either for a 'fish attracting device' or a 'floating attracting device'. Whichever case, these are structures that attract fish either because they provide shade or hiding places, or because they have lights that attract fish and squid.

Family selection: use as parents of certain families which are more ideal for a certain characteristic.

FAO: Food & Agriculture Organisation of the United Nations.

11

Fingerlings: young stages of fish (less than one year-old) that are used for culture in ponds or cages. Not a rigidly-applied term.

Fitness: in the genetic sense, having more offspring surviving to reproduce (=reproductive or Darwinian fitness).

Gadoid fish: the group of fish that includes cod, haddock, pollack, saithe and hake.

Genetic drift: changes in genetic make-up due to chance, which becomes far more prevalent when population size decreases.

Harmful algal bloom (HAB): blooms of species or strains of phytoplankton that produce toxins or are otherwise detrimental.

Holding facilities: common term used to describe a variety of indoor (e.g. tanks) and outdoor (e.g. ponds, cages) 'containers' to culture fish, molluscs or prawns.

Homeostasis: the active maintenance by an organism of a constant, balanced, internal environment.

Homologous chromosomes: similar chromosomes from mother and father, which pair during cells' duplication.

Hybrid vigour: where hybrids, usually between species, survive or grow better than their parents.

Hydrology: the understanding and study of water movement and water relationships.

IFFO: International Fishmeal & Fish Oil Organisation.

Inbreeding: strictly a reduction in reproductive fitness due to breeding of close relatives, but loosely used for the consequences of having small numbers of parents.

Interspecific: between two species.

Introgression: complete mixing of two species by random mating.

Larvae: developmental stages of an animal that differ from the adult in physical appearance and the type of food they eat.

Local adaptation: evolutionary mechanism which leads to optimal fitness in a certain locality.

Lymphocyte: a white blood cell, involved in the specific immune response.

Macrofauna: animals retained on sieve sizes between about 0.5 and 4 mm.

Mangroves: tropical/subtropical tidal salt marsh communities dominated by trees and shrubs and with tidal changes in water volume and salinity. Important nursery areas for fish and shellfish.

Mass selection: use of individuals that are superior for a certain characteristic as parents in breeding programme.

Methanogenesis: a fermentative process whereby both a highly reduced product (methane) and a highly oxidised product (carbon dioxide) are produced without the requirement for an external oxidant. This process only occurs in extremely reducing (organically rich) environments and may lead to the release of gas bubbles from sediment.

Nauplii: plural of nauplius. Early stages of the development of many types of crustaceans, including prawns and lobsters.

Neoplasm: an uncontrolled growth of cells; a cancer.

Nursery grounds or beds: natural areas of aquatic ecosystems such as seagrass beds, coral reefs and lagoons. They are protective habitats with high productivity that are used by young stages of aquatic animals such as prawns and fish.

Outbreeding depression: reduction in reproductive fitness of offspring of interbreeding between individuals from different locally-adapted groups.

Parasite: an organism which lives on, or in, another organism (the host), and obtains its nourishment at the host's expense.

Pathogen: an organism which induces a disease reaction in another organism.

Penaeids: common name for members of the crustacean family Penaeidae, made up of relatively large shrimps/prawns (N.B. the terms 'shrimp' and 'prawn' have no standard

11

meaning. Since larger members of this group are called prawns in the U.K., this colloquial term has been used throughout the book).

Phagocyte: a cell which can engulf foreign particles or organisms, inactivate them, and digest them.

Plankton: freshwater or marine plants (phytoplankton) or animals (zooplankton) living in the water column, mostly microscopic to pinhead sized, that have little or no control of their lateral position in the water (though zooplankton may migrate up and down considerable distances) – they drift with currents.

Population: a group of individuals within a species, presently occupying a distinct area and reproductively-isolated from other such groups.

Protist: primitive group of organisms, including protozoans and slime moulds.

Primary production: production of plant material by the process of photosynthesis.

Pseudofaeces: material filtered by bivalves that is not digested but is packaged and ejected.

Ranching: release of juvenile cultured animals to grow and feed naturally, usually in the ocean, with the aim of subsequent harvest of market size individuals.

Redox potential: a measure of the balance between reducing and oxidising processes. Where strongly negative, reducing processes dominate and where positive, towards the sediment surface, oxidising processes dominate. Measurement of Redox potential is used as a quick method of assessing the degree of sedimentary organic enrichment.

Resistance: the ability to resist disease.

Rickettsia: a parasitic, disease-producing organism, intermediate in size and complexity between a virus and a bacterium.

Salmonids: fish that include Atlantic and Pacific salmon as well as many types of trout.

Sea louse: colloquial (and undesirable) term for a range of species of external copepod parasites of fish. Copepods are small crustaceans, most of which are free-living. The parasitic species are mainly of the genera *Caligus* and *Lepeophtheirus*.

Spat: young stages of bivalve molluscs, in particular the stage at which the bivalve larvae stop living in the plankton and settle onto the surfaces of rocks, seaweed or adult bivalves.

Stocking: release of cultured aquatic animals into the wild.

Strain: a genetically distinct group of organisms in culture, usually derived from a particular population (see above).

Stress: an integrated physiological response of the body to stressful circumstances, which will result in the restoration of homeostasis, unless the stressful situation persists or is excessive, when disease or death may result.

Sulphide oxidising bacteria: these bacteria extract energy from the oxidation of sulphide in the presence of oxygen. White mats of these bacteria are found in sulphide rich sediments where reducing processes dominate to the sediment surface such that there is a good supply of both oxygen and sulphide.

Transgenic organism: an organism in which a gene from another species has been inserted into the genome.

Vaccination: administration of a vaccine (commercially produced antigens of a particular disease) which does not cause disease, but induces an immune response and resistance to the disease.

Virus: an ultramicroscopic parasite, which reproduces by infecting the cells of another organism.

12 Addresses of contributors

Dr Kenneth Black, Dunstaffnage Marine Laboratory, Oban, Argyll, PA34 4AD, Scotland, UK.
[kdb@dml.ac.uk]

Dr Gavin Burnell.
[g.burnell@ucc.ie]

Professor Tom Cross.
[t.cross@ucc.ie]

Dr Sarah Culloty.
[stzo8043@ucc.ie]

Professor John Davenport.
[jdavenport@zoology.ucc.ie]

Professor Maire Mulcahy.
[m.mulcahy@ucc.ie]

All at: Environmental Research Institute and Department of Zoology & Animal Ecology, University College Cork, Lee Maltings, Prospect Row, Cork, Ireland.

Professor Suki Ekaratne, Department of Zoology, University of Colombo, Colombo, Sri Lanka.
[postmast@zoo.cmb.ac.lk]

Professor Bob Furness, Institute of Biomedical and Life Sciences, University of Glasgow, Glasgow, G12 8QQ, UK.
[r.furness@bio.gla.ac.uk]

Dr Helmut Thetmeyer, Institut fuer Meereskunde Marine Oekologie Duesternbrooker Weg 20 D-24105 Kiel, Germany.
[hthetmeyer@ifm.uni-kiel.de]